Concrete Structures: Eurocode EC2 and BS 8110 Compared

Concrete Design and Construction Series

SERIES EDITORS

PROFESSOR F. K. KONG
Nanyang Technological University, Singapore

EMERITUS PROFESSOR R. H. EVANS CBE
University of Leeds

Concrete Structures: Eurocode EC2 and BS 8110 Compared

Edited by R. S. Narayanan

Longman
Scientific &
Technical

Longman Scientific & Technical
Longman Group UK Limited
Longman House, Burnt Mill, Harlow,
Essex, CM20 2JE, England
and Associated Companies throughout the world.

© Longman Group UK Limited 1994

First published 1994

1036720

British Library Cataloguing in Publication Data
A catalogue record for this book is available from the British Library

ISBN 0-582-06450-3

Set by 4Z in 10/12 pt Times
Printed and bound in Great Britain at the Bath Press, Avon.

CONCRETE STRUCTURES: EUROCODE EC2 AND BS 8110 COMPARED

Principal author

R. S. NARAYANAN   S. B. Tietz and Partners

With contributions from

PROFESSOR A. W. BEEBY   University of Leeds
R. T. WHITTLE   Ove Arup and Partners
K. R. WILSON   G. Maunsell and Partners

Contents

Acknowledgements

I am pleased to record my grateful thanks to many individuals and organisations without whose help it would have been difficult to produce such a valuable book.

First the three co-authors, Prof. Andrew Beeby, Robin Whittle and Keith Wilson for their dedicated work, fitting it in with their other busy schedules and putting up with my occasional harassing with good humour.

In the nature of the type of book we have produced, reference to the codes of practice is inevitable. However, reproduction of the codes has been kept to an absolute minimum and I gratefully acknowledge the permission given by the BSI in this regard. Complete copies of both EC2 and BS 8110 may be obtained from BSI, Linford Wood, Milton Keynes, MK14 6LE, UK.

I am also indebted to the DOE for their permission to draw upon the results of the parametric studies on the draft EC2 conducted by BCA on their behalf in 1989−1990.

A number of staff in my office helped with the production of the manuscripts and the partnership allowed the use of the computer facilities. My sincere thanks to all concerned.

R S Narayanan
Editor

1 Introduction

1.1 Background

Part 1 of Eurocode 2 (EC2) for the *Design of Concrete Structures* was published in the spring of 1992. Its full title is DD ENV 1992-1-1:1991. It comprises general rules and rules for buildings. The code is published by the national standards bodies of the member states on behalf of CEN (Comité Européen de Normalisation — the European standards body). The system adopted by CEN for the publication of codes and standards (Euronorms — ENs) involves the publication of the documents initially as pre-standards (ENVs). An ENV has a life of about three years, to enable member states to familiarize themselves with the new code. Normally at the end of this period the ENV will be amended in the light of any comments arising out of its use and will be published as an EN.

A clear understanding of the ENV will therefore be necessary to enable constructive comments to be made. One way of evaluating the ENV EC2 is by reference to the standard in current use, i.e. BS 8110. Comparison of the results of using the two codes will help form useful judgements. The aim of this book is to present such comparisons for all major aspects of the code. Parametric studies were conducted by BCA (British Cement Association) for the DoE (Department of the Environment), to evaluate all significant aspects of design. The results of these studies have been reproduced as relevant by kind permission of BRE (Building Research Establishment). Each chapter of this book starts with a general discussion of the treatment of the subject-matter and then goes on to provide comparisons.

1.2 Scope

The main chapter headings of ENV EC2 and BS 8110 are given under the headings below.

1.2.1 ENV EC2 — Part 1

Introduction, Basis of design, (fundamental requirements, definitions, and classifications, design requirements, durability, analysis), Material properties (concrete reinforcing steel, prestressing steel, prestressing devices), Section and member design (durability requirements, design data, ultimate limit studies, bending and longitudinal force, shear torsion punching, buckling, serviceability), Detailing, Construction and workmanship, Quality control, Appendices, Determination of the effects of time-dependent deformation of concrete, Non-linear analysis, Supplementary information on buckling, Checking deflections by calculation).

1.2.2   BS 8110

Part 1
General, Design objectives and general recommendations, Design and detailing reinforced concrete (design basis and strength of materials, structures and structural frames, concrete cover to reinforcement, beams, solid slab, ribbed slabs, flat slabs, columns, wall, staircases, bases, considerations affecting design details), Design and detailing prestressed concrete, Design and detailing precast and composite construction, Concrete: materials, Specification and construction, Specification and workmanship: prestressing tendons.

Part 2
General, Non-linear methods of analysis for the ultimate limit state, Serviceabililty calculations, Fire resistance, Additional considerations in the use of lightweight aggregate concrete, Autoclaved aerated concrete, Elastic deformation, creep drying shrinkage and thermal stress of concrete, Movement joints, Appraisal and testing of structures and components during construction.

As can be seen from the above, EC2 Part 1 is broadly comparable to BS 8110 Parts 1 and 2, although the BS covers more ground. While EC2 Part 1 does not deal with matters such as fire resistance and lightweight concrete, it must be remembered that EC2 is to be written in several parts as shown below:

Part 1.2 *Plain and Lightly Reinforced Concrete Structures*
Part 1.3 *Precast Concrete Structures*
Part 1.4 *The Use of Lightweight Aggregate Concrete*
Part 1.5 *The Use of Unbonded and External Prestressing Tendons*
Part 1.6 *Design for Fatigue of Concrete Structures*
Part 2   *Reinforced and Prestressed Concrete Bridge Structures*
Part 3   *Concrete Foundations and Piling*
Part 4   *Liquid-retaining Structures*
Part 5   *Temporary Structures; Structures with Short Design Life*
Part 6   *Massive Civil Engineering Structures*
Part 10 *Fire Resistance of Concrete Structures*

The above parts of EC2 are in varying stages of development. It will be a number of years before a full complement of Eurocodes is in place.

This is an appropriate juncture at which to introduce the role of the National Application Document (NAD), which is part of an ENV document. It is prepared by every member state for application within its jurisdiction. For example the

NAD is needed to point the way when a designer uses (say) ENV EC2, which makes reference to a document which has not been published even as an ENV. In the present context examples include loading code, fire design and reinforcement properties. The NAD is meant to provide operational guidance to enable the ENV EC2 to be used in the absence of a CEN standard to cover these matters. The UK NAD refers the designer to BS codes for loading, albeit with some modifications to preserve the compatibility between the different documents.

There is another important role for the NAD which is discussed in section 1.3 below.

1.3   Layout

When observing the contents list in section 1.2, one of the features which will be apparent is the way in which the chapters are ordered.

In BS 8110, there are sections dealing with beams, slabs, columns and so on, whereas EC2 has chapters on bending, shear, torsion, buckling, etc. Thus the Eurocodes by and large have arranged the chapters on the basis of phenomena whereas the BS uses element types.

In common with other Eurocodes, EC2 is set out with principles and application rules. The clauses which are principles are prefixed with P, e.g. clause 4.3.1.1 P(1). Principles comprise general statements, definitions, requirements and sometimes analytical models. No alternatives are permitted and all designs should comply with them. Application rules are rules that are generally adopted. These follow the principles and satisfy their requirements. The clauses which are application rules are indented and they do not (by contrast with principles) have any alphabetic prefix, e.g. clause 4.3.1.1 (5). Alternative rules may be used provided that it can be demonstrated that they comply with the principles.

In comparison to EC2, BS 8110 contains considerably more operational information (e.g. tables of bending moment coefficients for various elements). In many cases these can be used in conjunction with EC2, but any limitation in application should be recognized.

Some parameters in EC2 are designated by ⌴ , commonly referred to as boxed figures. The boxed values in the code are indicative only. The member states are required to fix the box values applicable within their jurisdiction. While it is hoped that a large number of the boxed values will be harmonized in due course, some differences are unavoidable to reflect the climatic differences and the long-standing engineering culture in the different countries.

Nomenclature in the two codes are not dissimilar in many instances. There are, however, some differences, but these are unlikely to present any problems of understanding, as the terms are fully defined. On the whole EC2 tries to use terms such that they cover a wide variety of situations. As a result some of these terms may appear unusual to the regular users of British codes. Thus 'loads' are referred to as 'actions'; 'bending moment', 'shear forces' and 'deflections' are 'action effects'; 'dead loads' are called 'permanent actions' and 'superimposed loads' are referred to as 'variable actions'.

2 Basis of Design

2.1 Symbols

Major symbols used in this chapter are given below.

2.1.1 EC2

$G_{k,\text{inf}}$ lower characteristic value of a permanent action
$G_{k,\text{sup}}$ upper characteristic value of a permanent action
G_k characteristic values of permanent actions
Q_{Ind} indirect variable action
a_{d} design values of geometrical data
a_{nom} nominal value of geometrical data
γ_G partial safety factor for permanent actions
γ_{A} partial safety factor for accidental design situations
γ_Q partial safety factor for any variable action
ψ factors defining representative values of variable actions
ψ_0 used for combination values
ψ_1 used for frequent values
ψ_2 used for quasi-permanent values

2.1.2 BS 8110

γ_{F} partial safety factor for load
γ_{m} partial safety factor for strength of materials
E_{n} nominal earth load
G_k characteristic dead load
Q_k characteristic imposed load
W_k characteristic wind load
f_{cu} characteristic strength of concrete
f_{y} characteristic strength of reinforcement
f_{pu} characteristic strength of a prestressing tendon

2.2   Relevant Chapters (Sections) in the Codes

The main chapters in the codes, where the relevant information is presented, are noted below.

2.2.1   EC2

2.0 Basis of design, 2.1 Fundamental requirements, 2.2 Definitions and classification, 2.3 Design requirements, 2.4 Durability, 3.0 Material properties, 3.1 Concrete, 3.2 Reinforcing steel, 3.3 Prestressing devices, 4.2 Design data, 4.2.1 Concrete, 4.2.2 Reinforced concrete, 4.2.3 Prestressed concrete.

2.2.2   BS 8110

2.0 Design objectives and general recommendations, 2.1 Basis of design, 2.2 Structural design, 2.3 Inspection of construction, 2.4 Loads and actual properties, 2.5 Design based on tests.

2.3   General

The principles in EC2 Part 1 are meant to be applicable to all structures. Therefore the character of the code is more general than it would need to be were it to apply to buildings only. On the other hand, BS 8110 is basically applicable to buildings. It is able to provide information directly and in a prescriptive form. Practising engineers naturally find the prescriptive method more attractive. Both codes use limit state principles.

While the basic aims are the same, there are differences of approach, particularly in the way loads are combined. This is discussed in the following sections.

2.4   Loads, Load Combinations and Partial Safety Factors

2.4.1   Loads

The magnitudes of loads to be used in design are based on characteristic values of loads in both codes.

For permanent loads EC2 draws a distinction between loads with small and large variations. Here the variation is considered small if the difference between the 95 percentile load and 5 percentile load is less than 20 per cent of the mean value. In such cases the mean value is used as the characteristic value. In the case of large variations, two characteristic values should be considered, namely upper ($G_{k,\text{sup}}$) and lower ($G_{k,\text{inf}}$). They correspond to the 95 and 5 percentile values. Such considerations may be relevant when dealing with, say, the weight of a slab or a wall cast against earth. BS 8110 does not make such an explicit distinction in the definition of the characteristic value of dead loads.

For variable actions the definition of the characteristic loads is the same in both the codes, generally 95 percentile loads in a reference period of 50 years (for wind loading it is 98 percentile load in one year). In many cases nominal values, often specified by the client, are treated as characteristic values.

Numerical values of loads will be given in EC1 (*Eurocode on Actions*). Until EC1 is published the UK NAD advises that BS codes such as BS 648, BS 6399 Parts 1 and 2 and CP3 Chapter V, Part 2 should be used with some modifications.

Loads vary in time and space. If this variation is fully known, it will be possible to utilize this in a probabilistic design. EC2 uses this approach. When one variable load is involved the design loading for ultimate limit state does not present any difficulties. It is

$$\gamma_G G_k + \gamma_Q Q_k$$

where γ_G is the partial safety factor for the permanent load, G_k the characteristic value of the permanent load, γ_Q the partial safety factor for the variable load and Q_k the characteristic value for the variable load. (Note that suffix k denotes characteristic values.)

When there are two variable actions (Q_1 and Q_2) of independent origin (there is no statistical correlation between the two loads), ($\gamma_{q1} Q_{k1} + \gamma_{Q2} Q_{k2}$) will not give the same probability of occurrence as $\gamma_Q Q_k$ in the previous example. To maintain the same probability of occurrence, joint probability will need to be considered. If the variation of Q_1 and Q_2 with respect to time is known, the realistic loading can be estimated. EC2 wants the designer to consider $\gamma_{Q1} Q_{k1} + \psi_{0,2} \gamma_{Q2} Q_{k2}$ and $\psi_{0,1} \gamma_{Q1} Q_{k1} + \gamma_{Q2} Q_{k2}$ and use the more critical of the two loads. The multiplier is the factor which provides a combination value. Its value depends upon the type of load and use of the structure (i.e. wind, snow, climatic, domestic or office use). The same logic applies when there are more than two variable loads. The principle can be understood as follows: when there are a number of independent variable actions, consider the action with the highest value as the dominant action. Combine the design value of the dominant, $\gamma_Q Q_{k,dom}$, with the sum of the combination values of all other actions, $\Sigma \psi_{0,i} \gamma_{Qi} Q_{ki}$. Where the dominant load is not obvious, each load should be treated in turn as dominant. Normally this is not necessary in practical situations.

In the accidental design situation, the load combination for ultimate loads may be expected to be different. The simplest way to approach this is to superimpose an accidental load on the service conditions of the structure. There may of course be more than one variable action involved in the service condition. Therefore the problem reduces to one of estimating realistic service loads at the time of occurrence of an accident. Knowledge of the variation of the load with time will be useful. Now 'the frequent' value of the dominant action is combined with the 'quasi-permanent' value of other actions. The frequent value is denoted as $\psi_1 Q_k$, and the quasi-permanent value is denoted as $\psi_2 Q_k$. The above principles are set out in tabular form in Table 2.1. The value of the partial safety factors γ_s are tabulated in Table 2.2. The values of ψ factors will be given in EC1 (*Eurocode*

Table 2.1 EC2: Basic combination of actions for ultimate limit state

Design situation	Permanent actions G_{d}	Variable actions Q_{d}		Accidental actions A_{d}
		Dominant	Others	
Persistent and transient	$\gamma_G G_k(\gamma_p P_k)$	$\gamma_{Q1} Q_{k1}$	$\gamma_{Qi}\psi_{0i} Q_{ki}$	
Accidental	$\gamma_{GA} G_k(\gamma_p P_k)$	$\psi_{11} Q_{k1}$	$\psi_{2i} Q_{ki}$	$\gamma_A A_k$ or A_{d}

Note: For accidental design situations the partial safety factors for permanent actions $\gamma_{GA} = 1.0$ may be used if not specified otherwise; for variable actions γ_{Qi} is equal to 1.0; for accidental actions $\gamma_A = 1.0$ if A_{d} is specified directly.

Table 2.2 EC2: Partial safety factors — ultimate limit state

Action	Symbols	Situations	
		Persistent/ transient	Accidental
Permanent actions caused by structural and non-structural components			
Ultimate limit state: static equilibrium			
Unfavourable	$\gamma_{G,\mathrm{sup}}$	1.10	1.00
Favourable	$\gamma_{G,\mathrm{inf}}$	0.90	1.00
Ultimate limit state: strength condition			
Unfavourable	$\gamma_{G,\mathrm{sup}}$	1.35	1.00
Favourable	$\gamma_{G,\mathrm{inf}}$	1.00	1.00
Variable actions, unfavourable	γ_Q	1.50	1.00
Accidental actions	γ_A		1.00

Table 2.3 EC2: ψ-values (UK NAD values)

Variable actions	ψ_0	ψ_1	ψ_2
Imposed loads			
Dwellings	0.5	0.4	0.2
Offices and stores	0.7	0.6	0.3
Parking	0.7	0.7	0.6
Wind loads	0.7	0.2	0
Snow loads	0.7	0.2	0

Note: For the purposes of EC2 these three categories of variable actions should be treated as separate and independent actions.

on Actions), but for the time being NAD provides the values which are reproduced as Table 2.3.

EC2 also permits a simplified approach for building structures as follows. In design situations with only one variable action, the design load to be considered is

$$\Sigma\gamma_{G,j} G_{k,j} + 1.5 Q_{k,1}$$

In design situations with two or more variable actions, the design load to be

Table 2.4 EC2: Simplified combination — rules

Load combination	Permanent load		Variable load imposed		Wind	Prestress
	Adverse	Beneficial	Adverse	Beneficial		
1 perm + imposed	1.35	1.00	1.50	0	—	1.00
2 perm + wind	1.35	1.00	—	—	1.50	1.00
3 perm + imposed + wind	1.35	1.00	1.35	0	1.35	1.00

Table 2.5 BS 8110: Partial safety factors for different load combinations

Load combination	Load type					
	Dead		Imposed			
	Adverse	Beneficial	Adverse	Beneficial	Earth and water pressure	Wind
1. Dead and imposed (and earth and water pressure)	1.4	1.0	1.6	0	1.4	—
2. Dead and wind (and earth and water pressure)	1.4	1.0	—	—	1.4	1.4
3. Dead and wind and imposed (and earth and water pressure)	1.2	1.2	1.2	1.2	1.2	1.2

considered is

$$\Sigma\gamma_{G,j}G_{k,j} + 1.35 \sum_{i \geq 1} Q_{k,i}$$

In this case the effects of the first combination above should also be considered for the dominant load and the more unfavourable effect should be used in design. Table 2.4 shows the partial safety factors to be used for typical load combinations in building structures, based on this simplified approach.

BS 8110: Part 1 provides the partial safety factors to be used for routine design of building structures. Table 2.5 summarizes this. This basically compares with the simplified approach in EC2. The one striking difference is the reduction in partial safety factor on permanent loads, when superimposed loads and wind loads are combined. There is no guidance when more than two variable loads are involved. While EC2 treats wind loading as any other variable action, i.e. γ factors are the same, BS 8110 reduces the γ factors for wind loads.

BS 8110: Part 2 suggests approaches other than the prescriptive format given in Part 1. Statistical methods (which will be largely similar to the Eurocode approach) and the use of reduced partial factors in conjunction with the worst credible values are recommended as possible alternatives.

For serviceability the loading which is considered depends on the effect that is to be checked. In EC2, if irreversible damage is to be checked, for example cracking of a brittle partition, the 'rare' combination of loads is suggested.

Table 2.6 EC2: Basic combination of actions for serviceability

Combination	Permanent actions G_d	Variable actions	
		Dominant	Others
Characteristic (rare)	$G_k(P_k)$	Q_{k1}	$\psi_{0i}Q_{ki}$
Frequent	$G_k(P_k)$	$\psi_{11}Q_{k1}$	$\psi_{2i}Q_{ki}$
Quasi-permanent	$G_k(P_k)$	$\psi_{21}Q_{k1}$	$\psi_{2i}Q_{ki}$

Cracking in concrete is normally checked using the 'frequent' combination. Long-term effects such as creep and settlement should normally be checked using the 'quasi-permanent' combination. The combinations are set out in Table 2.6.

Simplified combinations are also permitted for the serviceability limit state. These are as follows. In design situations with only one variable action the load to be used is

$$\Sigma G_{k,j} + Q_{k,1}(+ P)$$

In situations with two or more variable actions, the load to be considered is

$$\Sigma G_{k,j} + 0.9 \sum_{i \geq 1} Q_{k,i}(+ P)$$

or the first combination above for the dominant load whichever is more unfavourable.

In BS 8110, serviceability checks are normally carried out using a combination of unfactored loads (corresponds to the rare combination in EC2). Part 2 of BS 8110, however, permits the use of 'expected values' of loads. This is similar to the Eurocode method.

In summary, it may be stated that the basic approach to establishing the design load is similar in both the codes. EC2 provides an explicit framework for the ideas in BS 8110:Part 2.

2.5   Partial Safety Factors for Materials

The partial safety factor for materials (γ_m) is shown in Table 2.7 (for EC2) and Table 2.8 (for BS 8110). The partial safety factors for reinforcement are the same in both the codes. However, for concrete, EC2 uses 1.5 throughout, whereas BS 8110 uses different values for bending, shear and bond. EC2 allows the use

Table 2.7 EC2: Partial safety factors for materials

Load combination	Concrete γ_c	Reinforcement or prestressing steel γ_s
Combinations (1)−(3) in Table 2.1	1.5	1.15
Combination (4) in Table 2.1	1.3	1.0

Table 2.8 BS 8110 partial safety factors for materials

Reinforcement	1.15
Concrete in flexure or axial load	1.50
Shear strength without shear reinforcement	1.25
Bond strength	1.4
Others (e.g. bearing stress)	≥ 1.5

of values other than 1.5 depending on the quality control. This must be based on verifiable quantitative data.

2.6   Durability

It is not possible to produce a fully rigorous comparison between the durability requirements of BS 8110 and those in EC2. There are two principal reasons for this. Firstly, the influence on durability of the principal parameters such as cement content, water/cement ratio, concrete strength and cover is not generally agreed. While it is accepted universally that durability will be improved by increasing the cement content, cover or strength or by reducing the water/cement ratio, there is no generally accepted quantitative relation between these factors. Secondly, the definitions of exposure conditions in the two documents are different and are entirely qualitative. Thus only a subjective comparison of the two documents is possible. Comparisons will be attempted in this chapter, but the reader must bear in mind that these comparisons are, to a considerable degree, speculative.

In using EC2, it needs to be remembered that durability is not covered fully by EC2 and reference has to be made to ENV206, the standard for the specification, production and control of concrete. This document gives the required mix proportions for the various environments and EC2 gives the required minimum covers. It also needs to be noted that the NAD has modified the cover requirements given in Table 4.2 of EC2 for some exposure conditions. Attempts will be made to compare BS 8110 both with the clauses as drafted in EC2 and as modified by the NAD.

The general principles of the methods for design for durability in the two codes are the same and can be summarized as follows:

1. Establish the exposure class for the member being considered.
2. For the exposure conditions established in (1) above, establish the appropriate maximum water/cement ratio and minimum cement content or, alternatively, a 'deemed to satisfy' minimum strength grade.
3. Establish an appropriate minimum cover.

While the general procedure is the same in the two codes, there are significant differences in detail.

As has been mentioned, the definitions of the exposure conditions are different.

In BS 8110, a 'trade-off' is permitted between cover and concrete quality (a higher concrete quality can be balanced against a lower cover). In EC2, the

Table 2.9 Exposure class definitions from BS 8110

Environment	Exposure conditions
Mild	Concrete surfaces protected against weather or aggressive conditions
Moderate	Concrete surfaces sheltered from severe rain or freezing while wet
	Concrete subject to condensation
	Concrete surfaces continuously under water
	Concrete surfaces in contact with non-aggressive soils
Severe	Concrete surfaces exposed to severe rain, alternate wetting and drying or occasional freezing of severe condensation
Very severe	Concrete surfaces exposed to sea spray, de-icing salts, corrosive fumes or severe freezing conditions while wet
Extreme	Concrete surfaces exposed to abrasive action

Table 2.10 Definitions of environments in EC2 and ENV206

Exposure class	Examples of environmental conditions
1. Dry environment	Interior of dwelling or office
2a. Humid without frost	Interior with high humidity
	Exterior components
	Component in non-agressive soil or water
2b. With frost	Exterior exposed to frost
	Components in non-aggressive soils with frost
	Interior with high humidity with frost
3. Humid with frost and de-icing salts	Interior and exterior components exposed to frost and de-icing salts
4a. Sea-water environment without frost	Components completely or partially submerged in sea-water or in the splash zone
	Components in saturated salt air
4b. With frost	Components partially submerged in sea-water and exposed to frost
	Components in saturated salt air exposed to frost
5a.	Slightly aggressive chemical exposure
5b. Aggressive	Moderally aggressive chemical exposure
5c. Chemical exposure	Highly aggressive chemical exposure

Note: Class 5 exposure can occur in combination with any of the other exposure classes. ISO classifications of chemical aggressivity exist (see ISO 9690).

possibilities for 'trading off' are very much more limited, a 5 mm reduction in cover being permitted where the concrete cube strength exceeds $50 \, \text{N/mm}^2$. A 5 mm reduction in cover is also permitted for slabs. EC2 gives different minimum concrete grades for different classes of cement, but the same cement content and water/cement ratio. Different 'classes' of cement are covered differently in BS 8110 which implies that concretes having the same strength will have the same durability regardless of the nature of the cement [ordinary Portland cement + pfa (pulverized fuel ash) or ggbfs (ground granulated blast furnace slag) would constitute a different 'class' in this instance] but the achievement of this strength would require different cement contents and/or water/cement ratios.

Tables 2.9 and 2.10 give the definitions of exposure in the two codes and Tables 2.11, 2.12 and 2.13 summarize the cover and mix parameter requirements

Table 2.11 Durability provisions in BS 8110

Exposure class	Aggressivity factor	Nominal cover (mm)				
Mild	1	25	20	20	20	20
Moderate	2	—	35	30	25	20
Severe	3	—	—	40	30	25
Very severe	4	—	—	50	40	30
Extreme	—	—	—	—	60	50
Maximum water/cement ratio		0.65	0.60	0.55	0.50	0.45
Minimum cement content (kg)		275	300	325	350	400
Minimum grade		30	35	40	45	50

Note: Minimum grades may be reduced by 5 N/mm^2 if compliance with the mix proportion limits can be demonstrated for OPC concretes. For concretes with pfa or ggfbs the grades should be met. This may well require higher cement content.

Table 2.12 Durability provisions in EC2 using indicative values

Exposure class	Aggressivity factor	Nominal cover (mm)					
1	1.05	20	20	20	20	20	20
2a	2.35	—	25	25	25	25	25
2b	3.10	—	—	30	30	30	30
3	3.90	—	—	—	45	45	45
4a	3.70	—	—	—	—	45	45
4b	4.00	—	—	—	—	—	45
Minimum water/cement ratio		0.65	0.60	0.55	0.50	0.55	0.50
Minimum cement content		260	280	280	300	300	300
Minimum grade (32.5 cement)		25	30	37	45	37	45
Minimum grade (42.5 cement)		30	37	45	50	45	50

Note: (1) The cover may be reduced by 5 mm in slabs. (2) The cover may be reduced by 5 mm where the concrete grade exceeds 50. (3) The cover should be increased by 10 mm for prestressed concrete. (4) EC2 gives minimum covers. For easy comparison with BS 8110, a tolerance of 5 mm has been assumed in order to obtain nominal covers.

in the two codes. The 'aggressivity numbers' are an attempt to compare the environments. Their derivation will be covered in section 2.6.1.

2.6.1   Comparison of Definitions of Exposure

Since the definitions of exposure in both codes are entirely descriptive, it is only possible to attempt a somewhat subjective comparison. The method adopted has been borrowed from the field of job evaluation and analysis. The method, called 'Whole job ranking by paired comparisons', requires a job description to be produced for each of the jobs to be ranked. Every job description is then compared with every other job description. In each comparison, the job deemed to be 'worth more' is given two points and the other zero. In any case where no decision can be reached, each job scores one. When all the comparisons have been made, the total scores for each job are calculated and the jobs ranked according to these scores. Comparisons are carried out by a number of teams acting independently

Table 2.13 Durability provisions in EC2 using values from NAD

Exposure class	Aggressivity factor	Nominal cover (mm)					
1	1.05	20	20	20	20	20	20
2a	2.35	—	35	35	35	35	35
2b	3.10	—	—	35	35	35	35
3	3.90	—	—	—	40	40	40
4a	3.70	—	—	—	—	40	40
4b	4.00	—	—	—	—	—	40
Minimum water/cement ratio		0.65	0.60	0.55	0.50	0.55	0.50
Minimum cement content		260	280	280	300	300	300
Minimum grade		30	37	45	45	45	45

For notes see Table 2.12.

and the final ranking is obtained by averaging the scores from the various teams. Experience with the method shows that the ordering of the jobs thus obtained agrees very well with the intuitive feelings of 'fairness' of staff.

It seemed that, by substituting the exposure descriptions given in codes for job descriptions, the same method could be used for ranking environments. It was felt, however, that rather than just comparing EC2 with BS 8110, a more sensitive result would be obtained if the descriptions were 'diluted' by the addition of some descriptions from other codes. The total number of exposure descriptions was therefore made up to 20 by the addition of descriptions from CP 110, the CEB Model Code and the Australian code. At the time that this was done it did not prove possible to set up panels so the comparisons were carried out independently by four individuals, all of whom had been involved in work on durability. Table 2.14 gives the final averaged scores obtained for the various environments described in EC2 and BS 8110. It was not felt useful to include the extreme exposure in BS 8110 or exposure classes 5a, b and c in EC2 since these are all for special conditions.

These results were then normalized so that the aggressivity of the four exposure conditions in BS 8110 — mild, moderate, severe and very severe — corresponded respectively to 1, 2, 3 and 4. The resulting normalized relative aggressivity factors are given in Tables 2.11—2.13. It will be seen that the total range of the scales of exposure agree very closely in the two codes with 'mild' agreeing very closely with EC2 class 1 and 'very severe' agreeing closely with EC2 class 4b. Between

Table 2.14 Average scores for the exposure classes in EC2 and BS 8110

BS 8110		EC2	
Mild	25	1	27
Moderate	66	2a	74
Severe	88	2b	94
Very severe	146	3	139
		4a	129
		4b	145

these limits, however, there are considerable differences with 'severe' seeming
to be much the same as EC2 class 2. All participants in the comparisons judged
class 4a to be less severe than class 3. It will be seen that exposure classes 2b,
3, 4a and 4b in EC2 all fall within the 'very severe' class in BS 8110. Looking
at the actual gradings in Table 2.14 suggests that BS 8110 may provide a set of
definitions which divide up the possible range of practical environments in a more
uniform manner than does EC2. Other forms of comparison have also been tried
and have been found to give very similar results.

2.6.2   Comparison of Total 'Durability Package'

Having developed a comparison of the definitions of exposure, the problem is
now to obtain some idea of the relative durability of structures designed for the
same purpose to the two codes. This cannot be done by direct comparison of
Tables 2.11, 2.12 and 2.13 as there are too many variables involved. For example,
is a member with 45 mm cover, 0.55 water/cement ratio, a cement content of
$325 \, \text{kg/m}^3$ and a characteristic strength of $40 \, \text{N/mm}^2$ more or less durable than
one with 40 mm cover, 0.5 water/cement ratio, a cement content of $300 \, \text{kg/m}^3$
and a characteristic strength of $45 \, \text{N/mm}^2$? Without proposing some relationship
between these variables and durability, no meaningful comparison is possible.
Some such relationships will be given below which will be used for comparisons,
but it cannot be stated strongly enough that there exist no relationships at present
which would command anything approaching general acceptance. This is the
fundamental problem which faced both the drafters of BS 8110 and EC2.

Two hypotheses will be proposed which seem to provide a good fit to the
provisions in BS 8110. Since consideration of concrete quality in terms of mix
proportions and grade are seen as alternatives, relationships will be proposed (a)
as a function of mix proportions, and (b) as a function of concrete grade.

Dealing first with a relationship between cover and mix proportions, it is
suggested that durability will increase with increasing cover and will decrease
with increase in the free water in the concrete. Free water is assumed to be the
water put into the mix in excess of that required to hydrate the cement. Assuming
that a water/cement ratio of 0.25 is required to hydrate the cement, the free water
is given by

$$W_f = \text{free water} = \text{cement content} \times (\text{w/c ratio} - 0.25)$$

The durability of a member might now be assumed to be given by the relation

$$D = kC^a/W_f^b$$

where D is the required aggressivity number, C the cover and k, a and b are
constants.

The proposed relationship is reasonably consistent with the provisions in
BS 8110 if a is assumed to be 1.0 and b is assumed to be 2.0. Figure 2.1 shows
the aggressivity numbers plotted against C/W_f^2 estimated from the values in

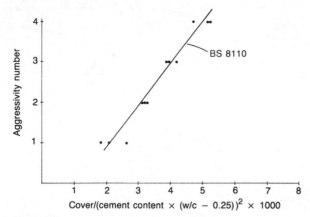

Figure 2.1

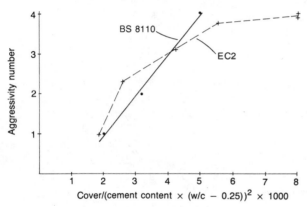

Figure 2.2

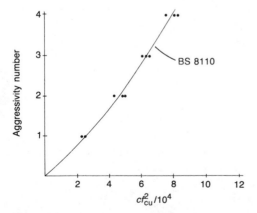

Figure 2.3

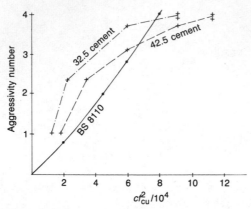

Figure 2.4

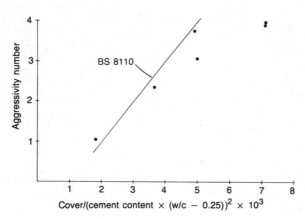

Figure 2.5 Dots correspond to EC2 results

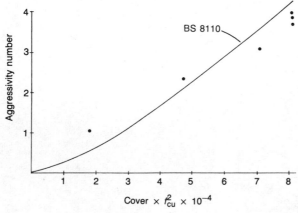

Figure 2.6 Dots correspond to EC2 results

Table 2.11. It will be seen that the BS 8110 provisions can be represented very closely by a single line. The scatter is no more than would be expected from the rounding of the cover figures to the nearest 5 mm, the cement contents to the nearest 25 kg and the water/cement ratios to the nearest 0.05 in the table.

The same process can be carried out for the values for 42.5 cement in Table 2.13. The result is shown in Fig. 2.2 which also shows the line obtained in Figure 2.1 for BS 8110. It will be seen that EC2 provides rather lower durability than BS 8110 below an aggressivity of 3, but considerably more durability above this value. Treatment of 32.5 cement is too speculative to be worth the attempt.

Dealing now with a relationship between durability, cover and concrete grade, we expect durability to improve with increase in cover and with increase in concrete strength. Thus a relationship of the following form seems reasonable:

$$D = kC^a f_{cu}^b$$

Figure 2.3 shows that the BS 8110 provisions are predicted well if a is taken as 1.0 and b as 2.0. Figure 2.4 shows the EC2 provisions plotted on this basis together with the line from Figure 2.3. BS 8110 proposes that other cements give the same durability as OPC for mixes having the same strength. The BS 8110 provisions therefore give the same line for both 42.5 and 32.5 cements. It will be seen that the conclusions drawn from Fig. 2.2 also hold for this alternative durability relation: EC2 will give less durable structures than BS 8110 for environments with low aggressivity numbers and more conservative structures for aggressivity numbers above 3. Assuming that the BS 8110 rules are right, EC2 may lead to very much lower durabilities where 32.5 cement is used. The conclusions from Figs 2.2 and 2.4 led to the NAD making significant changes to the indicative values given in EC2. Figures 2.5 and 2.6 show the relations used in Figs 2.2 and 2.4 but using the provisions from the NAD as summarized in Table 2.13 above. It will be seen that this gives fairly close agreement between EC2 and BS 8110. EC2, modified by the NAD, still gives a less flexible approach than BS 8110 since EC2 still does not permit the 'trade-off' between cover and concrete quality permitted in BS 8110.

3 Analysis

3.1 Symbols

Major symbols used in the two codes are listed below.

3.1.1 EC2

ΔH_j increase in the horizontal force on the floor, due to imperfections
ν angle of inclination assumed in the assessment of imperfection
δ ratio of redistributed moment to the moment before redistribution
x depth of the neutral axis at the ultimate limit state after redistribution

3.1.2 BS 8110

β_b ratio of the moment after redistribution to the moment before redistribution
x neutral axis depth

3.2 Relevant Chapters (Sections)

The main chapters (sections) where information or analysis is provided are listed below.

3.2.1 EC2

2.5 Analysis, 2.5.1 General provisions, 2.5.2 Idealization of the structure, 2.5.3 Calculation methods, Appendix 2 Non-linear analysis.

3.2.2 BS 8110

3.0 Design and detailing: concrete, 3.1 Design basis and strength of the materials, 3.2 Structures and structural frames, 3.4 Beams (3.4.1, 3.4.2, 3.4.3), 3.5 Solid slabs (3.5.1, 3.5.2, 3.5.3), 3.6 Ribbed slabs (3.6.2), 3.7 Flat slabs (3.7.1, 3.7.2, 3.7.3, 3.7.4), 3.8 Columns (3.8.1, 3.8.2).

3.3   General

In general, EC2 provides only the basic information required, whereas BS 8110 gives considerably more detailed information. Thus BS 8110 provides bending moment coefficients for a variety of structural elements, but EC2 expects the designer to obtain these from textbooks or manuals. EC2 gives hardly any guidance on the analysis of flat slabs, whereas BS 8110 devotes considerable space to this.

Both codes recognize all commonly used methods of analysis, for example elastic analysis with and without redistribution, plastic analysis and non-linear methods. Strut and tie models are also accepted in the two codes. Differences between the codes lie in the limitations they impose for each method.

Unlike BS 8110, EC2 draws a distinction between braced, unbraced sway and non-sway strutures. Thus a braced structure could still be a sway structure if the stiffness of the bracing elements is such that it leads to significant deformations causing further increase in bending moments in columns. EC2 provides a number of tests to check if a structure is a sway structure (Appendix 3). In BS 8110 it is implicitly assumed that all braced construction is also non-sway and all unbraced structures are sway structures. In practice this is probably true of real structures.

Unlike BS 8110, EC2 stipulates ductility characteristics for reinforcement as follows:

High ductility $\epsilon_{uk} > 5\%$ and $(f_t/f_y)_k > 1.08$
Normal ductility $\epsilon_{uk} > 2.5\%$ and $(f_t/f_y)_k > 1.05$

where ϵ_{uk} denotes the characteristic value of the elongation at maximum load, f_t the tensile strength of reinforcement and f_y the yield strength of reinforcement. This classification is relevant when plastic analysis is used and in determining the amount of redistribution of bending moments.

In EC2, design formulae are generally related to the cylinder strength f_{ck}, whereas BS 8110 uses cube strength f_{cu}. As an approximation the cylinder strength may be taken as 80 per cent of the cube strength.

3.4   Overall Stability

Perfection exists only in theory. Practical construction always has imperfections of varying degrees. The designer should recognize this and take steps to cater for the effects of the imperfections. Codes differ in the methods they choose to deal with this.

EC2 recommends that the structure should be assumed to have an inclination of ν to the vertical. For a structure with height ℓ

$$\nu = 1/100\sqrt{\ell}$$

Where n vertically continuous members act together

$$\nu = (\sqrt{(1 + 1/n)}/2)(1/100\sqrt{\ell})$$

The code also specifies minimum values of v. As the imperfection in every vertical member is unlikely to be the same, a reduction in the value of v is permitted depending on the number of vertically continuous members. The analysis may also be performed by assuming equivalent horizontal forces at each floor level equal to $v\Sigma V$ where ΣV is the sum of the vertical load at ultimate limit state at the floor being considered.

In BS 8110 the minimum horizontal load is taken as 1.5 per cent of the characteristic dead weight of the structure. The load is calculated for each floor level.

Table 3.1 shows the comparison of the minimum design horizontal forces according to the two codes. As can be seen, the BS invariably gives higher values.

Table 3.1 Comparison of minimum horizontal forces

	Ratio of horizontal force in EC2/BS 8110				
N	Q/G				
	0.25	0.5	1.0	1.5	2.0
2	0.40	0.49	0.67	0.83	1.0
≥ 4	0.28	0.35	0.475	0.60	0.725

Note: N is the number of storeys (storey height 4 m), *Q* the characteristic variable action and *G* the characteristic permanent action. This table is based on $v = 1/100\sqrt{\ell}$.

3.5   Load Arrangements for Continuous Frames

Both codes accept that sufficient number of load arrangements should be considered to calculate the critical design condition at all sections of the structure. For building structures the two codes recommend different load arrangements. For EC2:

1. Alternate spans carrying the design variable and permanent load; other spans carrying only the design permanent load.
2. Any two adjacent spans carrying the design variable and permanent loads; all other spans carrying only the design permanent load.

For BS 8110:

1. All spans loaded with maximum design ultimate load.
2. Alternate spans loaded with the maximum design ultimate load: all other spans loaded with the minimum design ultimate load.

In addition, for slabs BS 8110 also permits the use of a single load case of maximum design load on all spans, provided certain conditions are met.

3.6   Comparison of Design Bending Moments and Shears

A study was conducted by BCA for the DoE on beams supported on columns which are assumed to be fixed at their remote ends. The variables considered were as follows:

1. Number of spans (one to five equal spans);
2. Ratio of imposed to dead load (Q/G = 0.25, 0.5, 1.0, 1.5);
3. Ratio of column to beam stiffness (K_c/K_b = 0, 0.5, 1, 2);
4. Load patterns — as noted above and using γ_G = 1.35 and γ_Q = 1.5 for EC2.

The structure studied is shown in Fig. 3.1. The results are presented in Table 3.2. A summary of the results presented is as follows:

Beam moments

At first support	EC2 4–9 per cent below BS 8110 values
Near middle of first span	EC2 4–11 per cent below BS 8110 values
First interior support	Very similar ($\pm$ 5 per cent)
Near middle of other spans	EC2 12–20 per cent below BS 8110
Other supports	EC2 up to 15 per cent above BS 8110

Columns

Outer columns	EC2 4–9 per cent below BS 8110
Interior columns	EC2 up to 50 per cent below BS 8110

As the partial factors in EC2 are slightly lower than in BS 8110, it may be expected that the bending moments in EC2 are likely to be some 5 per cent below those which apply to BS 8110. The reasons for the differences, apart from the basic 5 per cent noted above lie in the interaction of two factors:

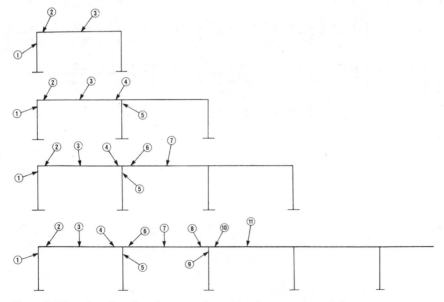

Figure 3.1 Location of sections for comparison of bending moments and shears

Table 3.2 Comparison of bending moments — beams/columns

		Ratios of EC2 moment/BS 8110 moment										
							Location					
K_c/K_b	No of spans	1	2	3	4	5	6	7	8	9	10	11
						$Q/G = 0.25$						
0.0	1	0.000	0.000	0.958								
0.0	2	0.000	0.000	0.870	0.958	0.000						
0.0	3	0.000	0.000	0.895	0.994	0.000	0.904	0.728				
0.0	5	0.000	0.000	0.888	0.980	0.000	0.980	0.771	1.030	0.000	1.030	0.922
0.5	1	0.958	0.958	0.958								
0.5	2	0.903	0.903	0.900	0.958	0.469						
0.5	3	0.906	0.906	0.910	0.977	0.689	0.990	0.820				
0.5	5	0.906	0.905	0.909	0.971	0.645	0.983	0.830	1.013	0.541	1.007	0.852
1.0	1	0.958	0.958	0.958								
1.0	2	0.916	0.916	0.913	0.958	0.469						
1.0	3	0.918	0.918	0.918	0.971	0.648	0.987	0.856				
1.0	5	0.918	0.918	0.914	0.967	0.635	0.981	0.860	1.002	0.513	0.396	0.369
2.0	1	0.958	0.958	0.958								
2.0	2	0.929	0.929	0.929	0.958	0.469						
2.0	3	0.930	0.930	0.930	0.963	0.600	0.983	0.892				
2.0	5	0.930	0.930	0.930	0.963	0.586	0.977	0.804	0.489	0.488	0.986	
						$Q/G = 0.5$						
0.0	1	0.000	0.000	0.955								
0.0	2	0.000	0.000	0.885	0.955	0.000						
0.0	3	0.000	0.000	0.904	1.012	0.000	1.012	0.783				
0.0	5	0.000	0.000	0.899	0.990	0.000	0.990	0.812	1.071	0.000	1.071	0.843
0.5	1	0.955	0.955	0.955								
0.5	2	0.910	0.910	0.908	0.955	0.625						
0.5	3	0.913	0.913	0.916	0.986	0.757	1.007	0.847				
0.5	5	0.913	0.913	0.915	0.975	0.729	0.994	0.860	1.044	0.666	1.034	0.070
1.0	1	0.955	0.955	0.955								
1.0	2	0.920	0.920	0.918	0.955	0.625						
1.0	3	0.921	0.921	0.922	0.976	0.731	1.001	0.874				
1.0	5	0.921	0.921	0.919	0.968	0.722	0.992	0.877	1.026	0.653	1.016	0.994
2.0	1	0.955	0.955	0.955								
2.0	2	0.931	0.931	0.931	0.955	0.625						
2.0	3	0.931	0.931	0.931	0.962	0.701	0.995	0.902				
2.0	5	0.931	0.931	0.931	0.962	0.692	0.986	0.903	1.004	0.636	1.000	

Table 3.2 cont'd

		Location										
K_c/K_b	No of spans	1	2	3	4	5	6	7	8	9	10	11
						$Q/G = 1$						
0.0	1	0.000	0.000	0.950								
0.0	2	0.000	0.000	0.901	0.950	0.000						
0.0	3	0.000	0.000	0.914	1.035	0.000	1.035	0.836				
0.0	5	0.000	0.000	0.910	1.002	0.000	1.002	0.854	1.121	0.000	1.121	0.977
0.5	1	0.950	0.950	0.950								
0.5	2	0.918	0.918	0.916	0.950	0.750						
0.5	3	0.920	0.920	0.922	0.996	0.821	1.027	0.876				
0.5	5	0.920	0.920	0.921	0.980	0.805	1.009	0.894	1.081	0.771	1.066	0.891
1.0	1	0.950	0.950	0.950								
1.0	2	0.925	0.925	0.923	0.950	0.750						
1.0	3	0.926	0.926	0.926	0.981	0.806	1.018	0.893				
1.0	5	0.926	0.926	0.924	0.970	0.801	1.006	0.895	1.055	0.764	1.040	0.900
2.0	1	0.950	0.950	0.950								
2.0	2	0.933	0.933	0.933	0.950	0.750						
2.0	3	0.933	0.933	0.933	0.961	0.661	1.009	0.912				
2.0	5	0.933	0.933	0.933	0.961	0.561	0.996	0.913	1.023	0.755	1.016	0.390
						$Q/G = 1.5$						
0.0	1	0.000	0.000	0.947								
0.0	2	0.000	0.000	0.909	0.947	0.000						
0.0	3	0.000	0.000	0.919	1.048	0.000	1.040	0.862				
0.0	5	0.000	0.000	0.916	1.009	0.000	1.009	0.875	1.150	0.000	1.150	0.891
0.5	1	0.947	0.947	0.947								
0.5	2	0.922	0.922	0.921	0.947	0.804						
0.5	3	0.924	0.924	0.925	1.002	0.851	1.038	0.891				
0.5	5	0.924	0.924	0.925	0.983	0.840	1.017	0.897	1.103	0.817	1.085	0.902
1.0	1	0.947	0.947	0.947								
1.0	2	0.928	0.928	0.926	0.947	0.804						
1.0	3	0.929	0.929	0.929	0.984	0.841	1.028	0.904				
1.0	5	0.929	0.929	0.927	0.972	0.838	1.013	0.905	1.072	0.813	1.054	0.909
2.0	1	0.947	0.947	0.947								
2.0	2	0.934	0.934	0.934	0.947	0.804						
2.0	3	0.934	0.934	0.934	0.960	0.830	1.017	0.918				
2.0	5	0.934	0.934	0.934	0.960	0.827	1.002	0.918	1.034	0.807	1.026	0.604

Note: Figures are ratios of EC2 values/BS 8100 values.

1. EC2 requires design for alternate and adjacent spans loaded while BS 8110 requires analysis for alternate spans loaded and all spans loaded. The use of alternate and adjacent spans loaded tends to give higher internal support moments relative to using alternate spans and all spans loaded.
2. EC2 uses 1.35 times the dead load on unloaded spans compared with 1.0 times the dead load used by BS 8110. This leads to a decrease in the maximum sagging moments in the spans produced by the 'alternate spans loaded' arrangement. In the columns, the EC2 approach greatly reduces the maximum 'out-of-balance' effect at columns and hence reduces the column moments very markedly.

Thus, EC2 gives an average 5 per cent reduction in beam moments, combined with a tendency for higher support moments and lower span moments. Internal column moments are substantially reduced. This last point is not as serious as might appear since the internal column moments tend to be relatively small in continuous beams with equal spans. BS 8110 currently allows such columns to be designed as nominally axially loaded (see BS 8110, clause 3.8.4.4).

Table 3.3 compares the EC2 pattern with that permitted in BS 8110 for slabs (single load case). The EC2 moments are from 5 to 3.0 per cent below the BS 8110 values in the span and 20−30 per cent higher over the supports. The ratios of the internal column moments should be considered as fairly meaningless since they are ratios of very small moments.

Table 3.4 compares the design shears generated by using EC2 with those from BS 8110 (arrangements (1) and (2) noted in section 3.5 above). The EC2 values lie generally within ±5 per cent of the BS 8110 values.

3.7   Points of Contraflexure

The position of the points of contraflexure (or, in some cases, where there are hogging moments over the whole span, in the size of this hogging moment) is affected by the choice of load patterns. Using a maximum and minimum dead load, as in BS 8110, causes the points of contraflexure in the unloaded spans to extend further into the span than with a uniform factor on the dead load on all spans. Figure 3.2 compares the extent of hogging moment predicted using BS 8110 to that given by the EC2 pattern. It will be seen that, while there are differences (about one-tenth of the span for situations with no columns but less with higher column stiffness), they do not appear very great. For beams, the practical significance of the differences is unlikely to be important as top steel will, in any case, be required over the whole span to support stirrups. Differences are more important in the case of slabs where it is more general practice not to carry top steel through areas not subject to hogging moments. Figure 3.3 compares the extent of hogging moment in slabs using the simplified load arrangement with those obtained using the EC2 pattern. Here, it will be seen that the EC2 is considerably more onerous than the BS 8110.

Table 3.3 Comparison of bending moments — slabs

		Ratios of EC2/BS 8110 pattern 2										
		Location										
K_c/K_b	No of spans	1	2	3	4	5	6	7	8	9	10	11
						Q/G = 0.5						
0.0	1	0.000	0.000	0.955								
0.0	2	0.000	0.000	0.884	1.194	0.000						
0.0	3	0.000	0.000	0.904	1.265	0.000	1.265	1.169	1.169			
0.0	5	0.000	0.000	0.903	1.238	0.000	1.238	0.869	1.339	0.000	1.339	0.925
0.5	1	0.955	1.194	0.909								
0.5	2	1.052	1.315	0.801	1.194							
0.5	3	0.938	1.308	0.819	1.233	1.892	1.233	0.989				
0.5	5	0.938	1.308	0.813	1.219	2.322	1.237	0.793	1.305	0.537	1.293	0.845
1.0	1	0.955	1.194	0.883								
1.0	2	1.028	1.285	0.764	1.194							
1.0	3	1.023	1.279	0.776	1.219	2.284	1.265	0.935				
1.0	5	1.023	1.279	0.776	1.210	2.447	2.240	0.757	1.283	7.773	1.270	0.806
2.0	1	0.955	1.194	0.856								
2.0	2	1.002	1.253	0.725	1.194							
2.0	3	1.002	1.253	0.731	1.202	3.048	1.244	0.886				
2.0	5	1.002	1.253	0.731	1.202	3.392	1.232	0.721	1.255	19.875	1.250	
						Q/G = 1.0						
0.0	1	0.000	0.000	0.95								
0.0	2	0.000	0.000	1.00	1.188	0.000						
0.0	3	0.000	0.000	0.986	1.294	0.000	1.294	1.393				
0.0	5	0.000	0.000	0.922	1.253	0.000	1.253	1.058	1.401	0.000	1.401	1.032
0.5	1	0.950	1.188	0.904								
0.5	2	1.093	1.367	0.882	1.188							
0.5	3	1.084	1.355	0.887	1.245	2.326	1.285	1.098				
0.5	5	1.084	1.355	0.882	1.225	2.949	1.261	0.904	1.351	7.485	1.333	0.915
1.0	1	0.950	1.188	0.879								
1.0	2	1.057	1.321	0.833	1.188							
1.0	3	1.052	1.315	0.836	1.226	2.899	1.273	1.008				
1.0	5	1.052	1.315	0.842	1.213	3.141	1.256	0.856	1.319	10.91	1.300	0.86
2.0	1	0.950	1.188	0.851								
2.0	2	1.022	1.275	0.779	1.188							
2.0	3	1.022	1.275	0.783	1.201	4.005	1.261	0.928				
2.0	5	1.022	1.275	0.783	1.201	4.511	1.245	0.794	1.279	27.96	1.27	

Note: Figures are ratios of EC2 values/BS 8110 values.

Table 3.4 Comparison of shear forces

K_c/K_b	No. of spans	Ratio of imposed to dead load = 0.25					Ratio of imposed to dead load = 0.5				
		2	4	6	8	10	2	4	6	8	10
0.0	1	0.958					0.955				
0.0	2	0.994	0.960				1.014	0.957			
0.0	3	0.984	0.965	0.904			0.997	0.966	1.009		
0.0	5	0.987	0.963	0.980	1.030	1.030	1.002	0.963	0.990	1.024	1.001
0.5	1	0.958					0.955				
0.5	2	0.985	0.957				0.998	0.953			
0.5	3	0.981	0.963	0.990			0.992	0.962	0.995		
0.5	5	0.983	0.977	0.983	1.013	1.007	0.995	0.960	0.985	0.999	0.989
1.0	1	0.958					0.955				
1.0	2	0.978	0.958				0.987	0.953			
1.0	3	0.978	0.961	0.979			0.987	0.959	0.989		
1.0	5	0.979	0.961	0.975	0.979	0.975	0.988	0.958	0.982	0.989	0.982
2.0	1	0.958					0.955				
2.0	2	0.972	0.958				0.978	0.954			
2.0	3	0.970	0.959	0.971			0.974	0.956	0.975		
2.0	5	0.970	0.959	0.969	0.969	0.967	0.974	0.956	0.971	0.972	0.968

K_c/K_b	No. of spans	Ratio of imposed to dead load = 1					Ratio of imposed to dead load = 1.5				
		2	4	6	8	10	2	4	6	8	10
0.0	1	0.950					0.947				
0.0	2	1.037	0.954				1.050	0.952			
0.0	3	1.012	0.967	1.030			1.021	0.967	1.042		
0.0	5	1.020	0.962	1.001	1.051	1.018	1.030	0.962	1.008	1.067	1.028
0.5	1	0.950					0.947				
0.5	2	1.014	0.947				1.023	0.944			
0.5	3	1.006	0.961	1.010			1.013	0.960	1.018		
0.5	5	1.009	0.958	0.995	1.016	1.000	1.018	0.957	1.000	1.025	1.007
1.0	1	0.950					0.947				
1.0	2	0.998	0.948				1.004	0.945			
1.0	3	0.997	0.957	1.000			1.003	0.956	1.007		
1.0	5	0.999	0.955	0.990	1.000	0.990	1.008	0.954	0.995	1.006	0.995
2.0	1	0.950					0.947				
2.0	2	0.984	0.949				0.987	0.946			
2.0	3	0.979	0.953	0.980			0.982	0.951	0.983		
2.0	5	0.979	0.953	0.975	0.975	0.970	0.982	0.951	0.977	0.977	0.971

Note: Figures are ratios of EC2 values/BS 8110 values.

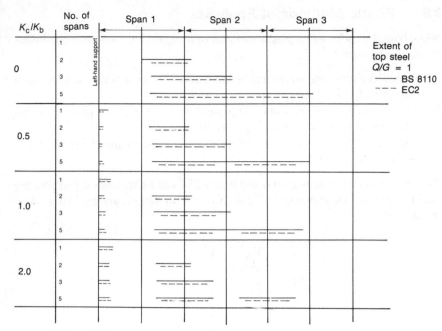

Figure 3.2 Extent of hogging moments — beams

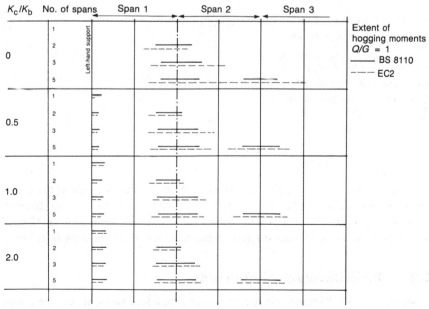

Figure 3.3 Extent of hogging moments — slabs

3.8   Plastic Methods of Analysis

While both the codes allow plastic methods of analysis, EC2 imposes conditions for its use, namely:

1.   Normal ductility steel should not be used, unless its application can be justified (presumably in relation to rotation capacity).
2.   The area of tension reinforcement should not exceed at any point or in any direction a value corresponding to $(x/d) = 0.25$.
3.   The ratio of the moments over continuous edges to the moments in span should lie between 0.5 and 2.0.

The limitation of the neutral axis depth to $0.25d$ can be translated into limiting steel percentages as shown in Fig. 3.4. No such limitations are explicitly stated in BS 8110.

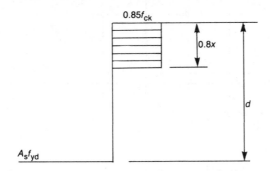

For plastic analysis x/d is limited to 0.25. Equating the internal forces, we get

$$0.8x(0.85f_{ck}/\gamma_c)b = A_sf_{yk}/\gamma_s$$

For

$$\gamma_c = 1.5 \quad \text{and} \quad \gamma_s = 1.15$$
$$A_s/bd = 0.130 \, (f_{ck}/f_{yk})$$

Limiting steel percentages for $f_{yk} = 460 \, \text{N/mm}^2$ are as follows:

f_{ck} (N/mm^2)	Steel percentage
20	0.565
25	0.707
30	0.848

Figure 3.4 Maximum steel for plastic analysis

In practice (at least in the UK) the plastic methods are popular only in the design of slabs. Here the neutral axis depth is generally low. However, the prohibition of normal ductility steel will present a problem, as small diameter rods (≤ 12 mm) are unlikely to comply with the requirements of the code for high ductility.

3.9   Redistribution of Moments

Both EC2 and BS 8110 permit redistribution of bending moments in continuous beams. It has not been considered helpful to carry out comparisons of redistributed moments obtained using the two documents as the procedure is an arbitrary one carried out in the particular way the designer chooses. All that will be presented here, therefore, is a comparison of the rules which govern what can be done.

In principle, the two sets of rules are the same: both require that the resulting redistributed diagrams remain in equilibirum with the applied design loads and both limit the possible amount of redistribution by a ductility check. Both codes also recognize that redistribution can result in parts of a beam being unsatisfactorily reinforced. The differences lie in the rules given to cover the ductility and detailing requirements in the two documents.

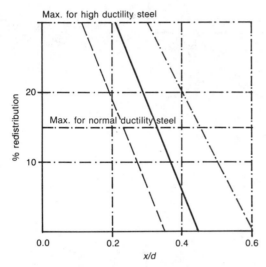

Figure 3.5 Redistribution of moment and limiting *x/d* values. (——) Concrete < C35/45, (--) concrete > C35/45, (-··-) BS 8110

The ductility limitations are summarized for the two codes in Fig. 3.5. In addition, EC2 defines the maximum permissible redistribution as a function of the ductility class of the reinforcement. A 30 per cent redistribution is permitted for high ductility steel, but only 15 per cent for normal ductility steel. It will be seen that the EC2 limits are considerably more severe than BS 8110 and somewhat more complicated. It should also be noted that EC2 does not permit any redistribution in sway frames, whereas up to 10 per cent redistribution is allowed in BS 8110.

There is some inconsistency between the rules controlling redistribution and the precondition for plastic analysis which amounts to infinite redistribution. As an example for C40/50 concrete, Fig. 3.5 will limit the depth of the neutral axis to $0.112d$. Nevertheless the neutral axis is allowed to go up to $0.25d$ if plastic analysis is performed.

The second point to note is that less redistribution is permitted with higher strength concrete which is considered to be more brittle. Thus, if a design using

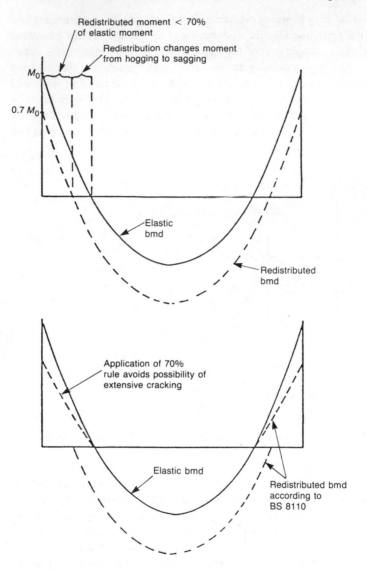

Figure 3.6 BS 8110: explanation of redistribution

C35/45 concrete just meets the limiting neutral axis depth for the amount of redistribution carried out, then, if the member was constructed using concrete of a higher characteristic strength than that specified, it would be unsafe! In practice, the higher strength concrete with a shallower neutral axis may well be adequate, without significantly affecting the required reinforcement.

BS 8110 introduces an additional limitation not explicitly given in EC2 to the effect that at no section throughout the beam should the redistributed moment be less than 70 per cent of the unredistributed ultimate bending moment. The purpose of this rule may be seen by reference to Fig. 3.6. It will be seen that, if the support moment is reduced by 30 per cent, the hogging moment at all points up to the point of contraflexure will have been reduced by more than 30 per cent and that over part of the area, the sign of the moment will have been reversed. The redistribution only occurs close to ultimate load when the support section yields; however, at lower loads before the redistribution occurs, there will be sections away from the support which will be overstressed and, at the least, serious cracking could result. The BS 8110 rule aims to produce a moment envelope which will avoid this possibility.

EC2 does not have a directly comparable rule but, in 2.5.3.4.2(7) it provides a restriction not in BS 8110. The objective of this provision would seem to be the more logical treatment of wide supports.

4 Design for the Ultimate Limit State

4.1 Design of Sections for Flexure and Axial Load

The methods given for section design in EC2 should be immediately familiar to anyone used to BS 8110. Indeed, the results obtained using EC2 will commonly be indistinguishable from those obtained using BS 8110. EC2 does, however, add some complications which sometimes make the calculations more complicated than is necessary.

The basic assumptions relating to section design are, in principle, identical with the exception of the treatment of the ultimate strains in sections where the whole section is subject to compressive stress. For these, the ultimate compressive strain is taken as 0.0035 for the situation where the neutral axis is at the least compressed face of the member, and 0.002 for a section subject to axial compression. For intermediate situations the ultimate strain is calculated by assuming a strain of 0.002 at a level three-sevenths of the height of the section from the most compressed face. This assumption is considerably more complex to apply than the equivalent assumption in BS 8110 where a constant limiting strain of 0.0035 is assumed.

EC2 permits the use of a stress—strain curve for the reinforcement which is identical to that in BS 8110. As an alternative, it also permits the use of a relationship with a sloping upper branch, which takes strain hardening into account. In this case it is necessary to introduce a limiting tensile strain of 0.01 in the reinforcement. In practice, the use of this more complex relationship gives minimal economic advantage, but may be necessary for some types of rigorous computer analysis where a completely horizontal top branch to the curve may give convergence problems.

The remaining factor which is required to calculate reinforcement areas to satisfy specified moments or moments combined with axial force, is the design stress—strain curve for the concrete. Both codes use a parabolic–rectangular diagram as their basic diagram. The EC2 diagram is slightly simpler to use than that in BS 8110 since it has a constant strain of 0.002 at the point where the parabola joins the plastic portion of the curve. The two curves are compared in Fig. 4.1 for concrete with a cube strength of $30 \, kN/mm^2$ or a cylinder strength of

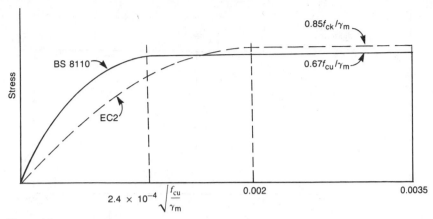

Figure 4.1

25 kN/mm². Comparisons for other concrete grades will be almost the same. It will be seen that the main difference is that the EC2 curve rises towards the plateau value less rapidly than does the BS 8110 curve, but that the plateau level is very similar. The fundamental parameters which arise from the stress–strain curve and which are needed in design are the average stress over the compression zone and the distance from the compression face to the centre of compression. These two factors are compared in Table 4.1.

Table 4.1 Comparison of EC2 and BS 8110. Parabolic–rectangular stress–strain curves

Grade	Average stress (N/mm^2)		Depth to centroid / Effective depth	
	EC2	BS 8110	EC2	BS 8110
20/25	9.2	10.1	0.416	0.456
25/30	11.5	12	0.416	0.452
30/37	13.8	14.6	0.416	0.447
35/45	16.1	17.6	0.416	0.442
40/50	18.4	19.4	0.416	0.439
45/55	20.6	21.2	0.416	0.436
50/60	22.9	22.9	0.416	0.434

Both EC2 and BS 8110 permit the use of simplified stress blocks. BS 8110 defines a rectangular stress block while EC2 permits the use of both a rectangular and a bilinear diagram. Figure 4.2 illustrates the possibilities. An approximate comparison of the three simplified diagrams can be given as follows:

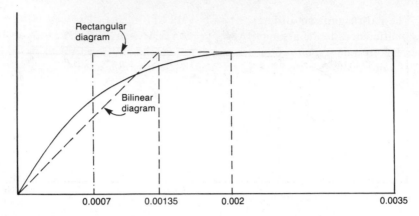

Figure 4.2

	Depth to centre of compression	Average stress (N/mm^2)
BS 8110 rectangular	$0.45d$	$0.4f_{cu}$
EC2 rectangular	$0.4d$	$0.37f_{cu}$
EC2 bilinear	$0.41d$	$0.37f_{cu}$

The approximation in the above figures is slight and arises because there is not an exactly constant ratio between the cube and cylinder strength.

The overall effect of these differences on the design of members subjected only to bending can best be seen by comparing the steel percentages calculated using these various assumptions for given values of M/bd^2. This is done in Table 4.2. It will be seen that the results given by all the methods are almost indistinguishable.

Table 4.2 Comparison of steel percentages calculated for $f_{cu} = 30$ and $f_y = 460$ using various stress blocks

M/bd^2	Values of $100A_s/bd$ for various stress blocks				
	BS 8110†	BS 8110‡	EC2§	EC2¶	EC2‖
0.5	0.13	0.13	0.13	0.13	0.13
1	0.26	0.26	0.26	0.26	0.26
1.5	0.4	0.4	0.4	0.4	0.4
2	0.54	0.54	0.54	0.54	0.54
2.5	0.7	0.69	0.69	0.69	0.69
3	0.86	0.85	0.86	0.85	0.86
4	1.23	1.19	1.21	1.2	1.21
5	1.67	1.59	1.64	1.62	1.63

† BS 8110 — parabolic–rectangular stress block.
‡ BS 8110 — rectangular stress block.
§ EC2 — parabolic–rectangular stress block.
¶ EC2 — rectangular stress block.
‖ EC2 — bilinear stress block.

The most significant difference between BS 8110 and EC2 arises, not from any differences in the assumptions for section behaviour, but as a result of the limitations imposed on neutral axis depth in order to ensure adequate ductility. The limits given in clause 2.5.3.4.2(3) of EC2 are as follows. For grade 35/45 or below:

$$x/d \leq 0.8\delta - 0.352, \text{ but in no case } > 0.45$$

For higher strength concretes:

$$x/d \leq 0.8\delta - 0.448, \text{ but in no case } > 0.35$$

In these relationships, δ is the ratio of the redistributed to the unredistributed moment (i.e. it is the same as β in BS 8110). There are additional limits to the maximum redistribution as a function of the type of reinforcement. The equivalent relationship in BS 8110 is

$$x/d \leq 0.4 + \beta$$

These neutral axis limits can be converted into maximum moments which sections can carry without the addition of compression steel. These limits, expressed in terms of maximum values of M/bd^2f_{cu}, are compared in Table 4.3 for various levels of redistribution. It will be seen from Table 4.3 that compression steel is required in designs to EC2 at much lower moments than in comparable designs to BS 8110.

Table 4.3 Comparison of maximum moment capacities of sections without compression steel

Redistribution (%)	Limiting values of M/bd^2f_{cu}		
		EC2	EC2
	BS 8110	$f_{cu} < =45$	$f_{cu} > 45$
0	0.16	0.14	0.114
10	0.16	0.119	0.092
15	0.148	0.108	0.08
20	0.136	0.097	0.068
25	0.122	0.085	0.054
30	0.107	0.073	0.041

The economic consequences of this may not be so significant as at first appears. This can be seen from an example. Consider a rectangular section designed to support a design value of M/bd^2 of 5 N/mm². Assuming that 20 per cent of redistribution has been carried out, the limiting values of M/bd^2 for EC2 and BS 8110 are respectively 2.91 and 4.08, so both codes require that compression steel be provided. EC2 requires 1.6 per cent of tension steel and 0.4 per cent of compression steel, giving a total steel percentage of 2 per cent. BS 8110 requires 1.5 per cent of tension steel and 0.25 per cent of compression steel, giving a total steel percentage of 1.75 per cent. Thus, though the EC2 limiting moment is 30 per cent below the BS 8110 value, the difference in total steel area is only 14 per cent. In practice, when bars have been chosen, the difference could be less.

Comparisons of the steel areas required to resist combined axial load and bending cannot be presented so conveniently, but the two codes will, in fact, give almost indistinguishable results. This can be seen for a typical case in Fig. 4.3 which superimposes design charts for the two codes for a rectangular section. There is a small element of approximation in the curve for BS 8110 since the shape of the BS 8110 parabolic—rectangular stress block changes slightly with concrete strength (see Table 4.1). The curve will be exact for a cube strength of 30 N/mm^2 and marginally wrong for higher strengths.

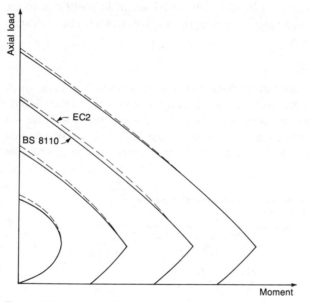

Figure 4.3

4.2   Shear

The basic principles behind the methods for the design for shear are the same in the two codes. These can be summarized as follows:

1.   If $v < v_c$ only nominal steel is required in beams. No shear reinforcement is required in slabs.
2.   If $v_c < v < v_{max}$ designed shear reinforcement is required.
3.   If $v > v_{max}$ section cannot be designed to support the required shear and a larger section must be chosen.

The notation in EC2 differs from that in BS 8110 and in comparing the clauses it needs to be remembered that:

V_{Rd1} in EC2 is equivalent to $v_c bd$ in BS 8110;
V_{Rd2} in EC2 is equivalent to $v_{max} bd$ in BS 8110;
V_{Rd3} in EC2 is the shear strength of a section which includes shear reinforcement.

However, while the basic principles do not differ, the calculation for each of the quantities mentioned above differs very considerably. Each item will now be considered in turn.

4.2.1   Strength of Sections Without Shear Reinforcement (v_c or V_{Rd1})

In both codes the strength of a section without shear reinforcement depends upon the concrete strength, the tensile steel percentage, the section depth and the axial force. Ignoring axial force for the present, the formulae in the two codes are as follows. In these equations the notation has been converted to BS 8110 notation.

1.   BS 8110:

$$v_c = 0.79(pf_{cu})^{1/3}(400/d)^{1/4}/\gamma_m$$

2.   EC2:

$$v_c = 0.0525f_{ck}^{2/3}(1.2+0.4\rho)(1.6-d/1000)/\gamma_m$$

In the above equations, $\rho = 100A_s/b_v d$.

Limitations are applied to the values of some of the variables in these equations. In BS 8110, ρ does not take a value >3 and d does not take a value >400 mm. Additionally, f_{cu} should not be taken as >40 N/mm². In EC2, ρ does not take a value >2 and d does not take a value >600 mm. There is no limit on the concrete strength in EC2. The values given for γ_m are 1.25 and 1.5 in BS 8110 and EC2 respectively.

Because of the number of variables involved, it is difficult to present a simple comparison between the predictions of the two codes. However, Tables 4.4a and b provide a comparison for concrete of 30 N/mm² cube strength in the format used in BS 8110. Table 4.5 gives correction factors by which the values in Tables 4.4a and b can be multiplied to adjust for other concrete strengths.

It will be seen that, except for very low steel percentages, EC2 generally gives lower values of v_c than BS 8110 for a concrete strength of 30 N/mm². This arises mostly from the lower value of the partial safety factor used in BS 8110. However, because EC2 gives a much greater influence to concrete strength than does BS 8110, the opposite may be true for higher concrete strengths. The range of ratios of v_c calculated according to EC2 to that calculated according to BS 8110 is roughly between 0.7 and 1.7. This may seem very large but the practical significance of the differences should not be overstated. Beams will always contain at least nominal shear reinforcement and this will tend to reduce the real differences between designs to the two codes. One-way and two-way spanning slabs supported

Table 4.4a   Values of v_c according to BS 8110

Steel (%)	Effective depth (mm)					
	150	200	250	300	400	600
0.15	0.46	0.42	0.40	0.38	0.36	0.36
0.30	0.57	0.53	0.51	0.48	0.45	0.45
0.50	0.68	0.63	0.60	0.57	0.53	0.53
0.75	0.78	0.73	0.69	0.66	0.61	0.61
1.00	0.86	0.80	0.76	0.72	0.67	0.67
1.50	0.98	0.91	0.86	0.83	0.77	0.77
2.00	1.08	1.01	0.95	0.91	0.85	0.85
3.00	1.24	1.15	1.09	1.04	0.97	0.97

Table 4.4b   Values of v_c according to EC2

Steel (%)	Effective depth (mm)					
	150	200	250	300	400	600
0.15	0.55	0.53	0.51	0.49	0.45	0.38
0.30	0.57	0.55	0.53	0.51	0.47	0.40
0.50	0.61	0.59	0.57	0.54	0.50	0.42
0.75	0.65	0.63	0.61	0.58	0.54	0.45
1.00	0.69	0.67	0.65	0.62	0.57	0.48
1.50	0.78	0.75	0.73	0.70	0.65	0.54
2.00	0.87	0.84	0.81	0.78	0.72	0.60
3.00	0.87	0.84	0.81	0.78	0.72	0.6

Table 4.5   Correction factors for different concrete strengths

	25	30	35	40	45	50
BS 8110	0.94	1.00	1.05	1.10	1.10	1.10
EC2	0.89	1.00	1.11	1.21	1.31	1.41

on beams are unlikely to require shear reinforcement whichever code is used. It will be seen later that the situation where the differences in v_c do have a major practical effect is in the design of flat slabs for punching shear.

The influence of the presence of axial force on shear is taken into account for reinforced concrete in BS 8110 and for both reinforced and prestressed concrete in EC2 by adding to v_c or V_{Rd1} an amount proportional to the axial stress. BS 8110 uses a more complex formulation for prestressed concrete which will be covered elsewhere in this book. The two formulae for the effect of axial load are as given below.

$$v_c' = v_c + 0.75NVd/A_cM \quad \text{(BS 8110)}$$

$$v_c' = v_c + 0.15N/A_c \quad \text{(EC2)}$$

It will be seen that the BS 8110 additional term is $5Vd/M$ times the EC2 term. Figure 4.4 illustrates how the terms differ over the span of a uniformly loaded

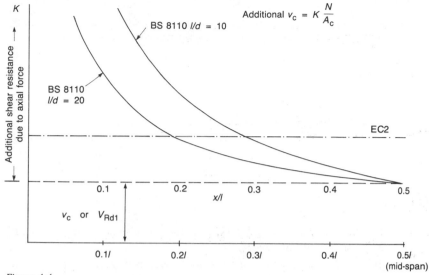

Figure 4.4

simply supported beam. BS 8110 will give rather higher shear strengths than EC2 close to supports, but EC2 is likely to give higher strengths in the centre half of the beam. The comparison of the effects of axial load are further complicated closer to the supports than two effective depths due to the slightly different treatments of the enhancement of strength close to supports in the two codes. This will be considered later in this chapter.

4.2.2   Strength of Sections with Shear Reinforcement

EC2 gives two methods of assessing the shear strength of sections reinforced in shear (or, if you like, of calculating the required areas of shear reinforcement). The first, or 'standard method' is very similar to that in BS 8110. The second, or 'variable strut inclination method' is similar in principle but gives the designer considerable scope to optimize the design. In some cases this second method can lead to considerable economies in shear reinforcement over the standard method and over BS 8110. The methods in BS 8110 and EC2 are all based on the assumption that shear reinforcement acts as tension elements in a virtual truss, as illustrated in Fig. 4.5. Analysis of this truss gives the shear reinforcement required to carry a shear force of V_s as

$$A_{sv} = \frac{V_s s \gamma_m}{z f_{yv}(\cot \theta + \cot \alpha) \sin \alpha}$$

In this equation A_{sv} is the area of one element of the shear reinforcement, s the spacing of the shear reinforcement, z the lever arm of the internal forces, f_{yv} the strength of the shear reinforcement, θ the angle between the axis of the beam

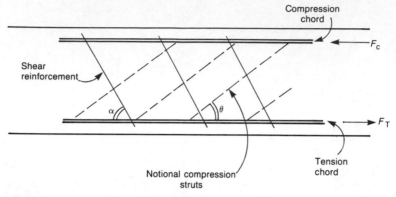

Figure 4.5

and the notional compressive struts and α is the angle between the axis of the beam and the shear reinforcement.

In the BS 8110 method and in the standard method in EC2 the angle of the notional compression struts, θ, is fixed at 45°. In addition, it is assumed that it is not necessary for the reinforcement to carry all the shear but only the shear in excess of that which can be carried by the concrete (i.e. $v - v_c$). The only difference between the two codes is that EC2 assumes a value of $0.9d$ for the lever arm (z in the above equation) while BS 8110 assumes a value of d. This leads to the following equations for vertical stirrups in the two codes:

$$\frac{A_{sv}}{b_v s_v} = \frac{(v - v_c)}{0.87 f_{yv}} \quad \text{(BS 8110)}$$

$$\frac{A_{sv}}{b_v s_v} = \frac{(v - v_c)}{0.78 f_{yv}} \quad \text{(EC2)}$$

Since v_c is likely to be slightly smaller in EC2 than in BS 8110, EC2 is likely to require about 15 per cent more stirrups than BS 8110 in most cases where the standard method is used.

In the variable strut inclination method, the angle θ is not fixed at 45° but can take values within the ranges

$$0.4 < \cot \theta < 2.5$$

for beams with constant longitudinal reinforcement and

$$0.5 < \cot \theta < 2.0$$

for beams with curtailed longitudinal reinforcement.

The choice of the angle θ is not entirely free as it will be found to be constrained by the rules for v_{max}, which is a function of θ for high shears. Also, unlike the standard method, it is assumed that, once shear reinforcement is required, all the shear has to be carried by the shear reinforcement. The equation for vertical stirrups using the variable strut inclination method thus becomes

$$\frac{A_{sv}}{b_v s_v} = \frac{v}{0.78 f_{yv} \cot \theta}$$

The approach envisaged for design is that, firstly, the value of $\cot \theta$ is calculated which gives $v = v_{max}$. If this value of $\cot \theta$ is greater than the maximum limiting value then the maximum value is taken. Otherwise the value of $\cot \theta$ corresponding to $v = v_{max}$ is used. The shear reinforcement is then calculated from the above equation.

4.2.3   Maximum Shear that can be Carried by a Section

The maximum shear that can be carried by a section is limited by the strength of the virtual compression strut of the assumed truss system. Both BS 8110 and EC2 provide rules governing this. Clearly, the force carried by the compression strut will depend upon the angle of inclination of the strut. Thus, for the BS 8110 method and the standard method in EC2, a value can be expected which is a function only of the concrete strength since the strut angle is assumed to be constant in both methods. However, a value which is also a function of the strut angle must be expected for the variable strut angle method.

The equations given for the maximum shear stress in the two codes are as follows:

$$v_{max} = 0.8\sqrt{f_{cu}} \le 5 \, \text{N/mm}^2 \quad \text{(BS 8110)}$$

$$v_{max} = f_{cd}(z/d)v(\cot \theta + \tan \theta) \quad \text{(EC2)}$$

where $v = 0.7 - (f_{ck}/200) \ge 0.5$.

To convert the EC2 formulae to be more easily comparable with BS 8110, it can be assumed that the cylinder strength of the concrete is 0.8 times the cube strength without significant error. Additionally, EC2 states that the lever arm, z, may be assumed to be $0.9d$ and f_{cd} is equal to $f_{ck}/1.5$. Remembering that $\theta = 45°$ for the standard method, the EC2 equations can now be rewritten as follows. For the standard method:

$$v_{max} = 0.24 v f_{cu}$$

For the variable truss angle method:

$$v_{max} = 0.48 v f_{cu}/(\cot \theta + \tan \theta)$$

In both these equations:

$$v = 0.7 - f_{cu}/250 \ge 0.5$$

While EC2 omits to say so, θ should not be taken as greater than 45° when calculating v_{max} (i.e. the strut should not be assumed to be nearer to the vertical than 45°).

Table 4.6 compares the values of v_{max} given by the various formulae.

Table 4.6 Comparison of maximum allowable shear stresses

Cube strength (N/mm^2)	BS 8110	EC2		
		$\theta = 27°$	$\theta = 35°$	Standard
25	4.00	2.91	3.38	3.60
30	4.38	3.38	3.92	4.18
35	4.73	3.81	4.42	4.70
40	5.00	4.19	4.87	5.18
45	5.00	4.54	5.28	5.62
50	5.00	4.85	5.64	6.00
55	5.00	5.13	5.95	6.34
60	5.00	5.36	6.22	6.62

It will be seen that EC2 permits significantly higher shears to be carried by sections than does BS 8110. There is, however, an additional factor in EC2 which may need to be taken into account in situations where axial forces are present (this includes prestressed members). EC2 requires a reduction in v_{max} where significant axial forces are present (Table 4.7). This is given in clause 4.3.2.2(4), but can be presented in an alternative form as shown below:

$$v_{max.red} = kv_{max}$$

$$k = 1.67 + 1.63\rho'f_y/f_{cu} - 1.875s_d/f_{cu} \leq 1$$

where ρ' is the compression reinforcement ratio (A_{sc}/A_c) and s_d the design axial stress (Ns_d/A_c). It will be seen that it is only at very high levels of axial load that EC2 requires any reduction in v_{max}.

Table 4.7 Reduction factors to take account of the effect of axial forces on the maximum shear capacity in EC2

$A_{sc}f_y/A_cf_{cu}$	Ns_d/A_c		
	0.4	0.5	0.6
0.05	1.00	0.81	0.63
0.10	1.00	0.90	0.71
0.15	1.00	0.98	0.79
0.20	1.00	1.00	0.87
0.25	1.00	1.00	0.95
0.30	1.00	1.00	1.00

4.2.4   Shear Enhancement Near Supports

Both codes allow higher shears to be carried at sections close to supports. This effect is experimentally well proved and may be considered to occur because a proportion of the load is transmitted to the support by direct compression rather than by shear. In both codes the enhancement is expressed as a function of the parameter a_v/d where a_v is the distance from the face of the support to the section considered. In BS 8110, the shear that can be carried by the concrete section without shear reinforcement, v_c, may be increased by a factor $2d/a_v$, where a_v is less than $2d$. In EC2, the shear that can be supported without shear reinforcement is increased by a factor $2.5d/a_v$ where a_v is less than $2.5d$. Thus EC2 gives a slightly greater enhancement of capacity than does BS 8110. There is a further significant difference in principle between the way the two codes approach the strength enhancement near supports. EC2 states that the enhancement can be used to support *loads* which are within $2.5d$ of the support, while BS 8110 states that the enhancement may be used in the design of *sections* which are within $2d$ of the support. The BS 8110 method is thus clearly intended to be used for sections close to the supports even where the loads producing the shears are more distant from the support. It would seem, however, that the two codes probably have the same idea in mind since both BS 8110 and EC2 allow account to be taken of the enhancement for all types of loading in a simplified way. This is done by assuming the critical section for shear to be a distance d from the face of the support rather than at the face of the support. In EC2 this provision is given in clause 4.3.2.2(10) which, effectively, contradicts the first sentence of clause 4.3.2.2(9).

4.2.5 Additional Tensile Force in the Longitudinal Reinforcement

If the truss illustrated in Fig. 4.5 is analysed, it will be found that the bottom tension member is required to carry a tensile force of

$$T = M/z + 1/2V \cot \theta$$

BS 8110 and, indeed, many other current codes, ignore the second term and simply design for the bending term. EC2 does not and there is a requirement that it should be taken into account (this requirement is given in clause 4.3.2.4.4(5) in a more general form). This may either be done explicitly or by displacing the M/z curve an amount $\frac{1}{2} z \cot \theta$. This is called the 'shift rule' and will be mentioned again later in Chapter 6.

The EC2 provisions seem entirely logical in this case, but mean that, where reinforcement is curtailed, EC2 requires the curtailed bars to extend $0.45d$ for the standard method and up to $0.9d$ for the variable strut angle method, further than would be required by BS 8110.

4.2.6 Minimum Percentages of Shear Reinforcement

The minimum percentages of shear reinforcement are specified in a rather different way in the two codes. In BS 8110, minimum links are required to correspond to a design shear stress of 0.4 N/mm^2. This is given in Table 3.8 of BS 8110. EC2 gives minimum reinforcement ratios for shear in Chapter 5. Table 5.5 in EC2 gives shear reinforcement ratios as a function of reinforcement class and concrete grade. Table 4.8 compares the minimum reinforcement ratios specified

Table 4.8 Comparison of minimum shear reinforcement provisions

Code	f_{cu}	$f_y=200$	$f_y=400$	$f_y=460$	$f_y=500$
EC2	15–25	0.16	0.09	0.08	0.07
	30–45	0.24	0.13	0.12	0.11
	50–60	0.30	0.16	0.15	0.13
BS 8110	All	0.21	0.11	0.10	0.09

by the two codes. The basic point to note is that EC2 will require higher amounts of minimum reinforcement than BS 8110 for higher concrete strengths.

4.2.7   Spacing of Links

The maximum spacing of links is given in BS 8110 in clause 3.4.5.5 as $0.75d$ in the direction of the span, while at right angles to the span the legs should not be more than d apart and no longitudinal bar should be more than 150 mm from a vertical leg. The rules in EC2 are more complex and are given in clauses 5.4.2.2(7) and (9), the maximum spacings both longitudinally and laterally are a function of the applied shear. The comparable rules are given in Table 4.9.

Table 4.9   Minimum spacings of links in EC2

	Spacing (mm)	
	Longitudinal	Lateral
$v < v_{max}/5$	$0.8d \leq 300$	$d \quad \leq 800$
$v_{max}/5 < v < 2v_{max}/3$	$0.6d \leq 300$	$0.6d \leq 300$
$v > 2v_{max}/3$	$0.3d \leq 200$	$0.3d \leq 200$

Example 4.1

Design a rectangular section of breadth 300 mm and with an effective depth of 400 mm to withstand a design shear of 200 kN. The concrete strength is C30/37, the shear reinforcement has a characteristic strength of 460 N/mm^2 and the percentage of tension reinforcement is 1.2 per cent.

Design to BS 8110

The shear stress is given by

$$v = \frac{200 \times 1000}{400 \times 300} = 1.667$$

The shear stress that can be carried by the concrete alone is given by Table 3.9 in BS 8110 as 0.765.

$$(v - v_c) = 1.667 - 0.765 = 0.9.$$

This is more than 0.4 hence designed shear reinforcement is required.

$$100A_{sv}/b_v s_v = 90/0.87 \times 460 = 0.225$$

$$v_{max} = 0.8\sqrt{37} = 4.86$$

This is > 1.667 therefore the section is adequate.

Design to EC2 Standard Method

The design shear, $V_{Sd} = 200\,000$. The shear that can be carried by the concrete alone V_{Rd1} is given by Eq. 4.18 in EC2 as 82 000 N. This is less than the design shear so designed shear reinforcement is required. From Eq. 4.23 in EC2,

$$100A_{sv}/b_v s_v = \frac{(200\,000 - 82\,000) \times 100}{0.78 \times 460 \times 300 \times 400} = 0.274$$

The maximum shear that the section can carry is given by

$$V_{Rd2} = 0.5 \times 0.55 \times 30/1.5 \times 300 \times 0.9 \times 400 = 594\,000$$

This is $>200\,000$ therefore the section is adequate.

Design by the Variable Truss Angle Method

As for the standard method, $V_{Sd} = 200\,000$ and $V_{Rd1} = 82\,000$ hence shear reinforcement is required. Cot θ should be chosen at the most economical value of 2.5 unless this leads to too low a value of V_{Rd2}.

$$V_{Rd2} = 0.55 \times 300 \times 0.9 \times 400 \times 30/1.5 \times 2.5/(1+2.5^2)$$
$$= 409\,000$$

This is $>200\,000$ therefore cot θ may be taken as 2.5.

$$100A_{sv}/b_v s_v = 20\,000\,000/(0.9 \times 400 \times 2.5 \times 0.87 \times 460 \times 300)$$
$$= 0.185$$

4.2.8   Summary

It will be seen that all the calculations are about equally simple, but that the amounts of shear steel required differs very substantially between the three methods. Summarizing, the values of $100A_{sv}/b_v s_v$ obtained are as follows:

BS 8110	0.225
EC2 standard method	0.272
EC2 variable strut	0.185

It will be seen that the BS 8110 method lies almost exactly halfway between the two EC2 methods. It will also be seen that very substantial savings in shear reinforcement can be made by using the variable truss angle method.

Table 4.10 Comparison of values of $100A_{sv}/b_v s_v$

Shear stress	BS 8110	Eurocode		
		Standard	Variable θ	
1.00	0.10	0.12	0.12	Minimum values
2.00	0.29	0.38	0.22	
3.00	0.54	0.66	0.33	
4.00	0.79	0.94	0.53	
5.00	1.04	1.21	1.08	
6.00	—	—	—	Above v_{max}

Table 4.10 gives a comparison between the shear reinforcement requirements predicted by BS 8110 and by the two EC2 methods for a typical case. In producing Table 4.10 it has been assumed that the concrete cube strength was 40 N/mm^2 and that the tension steel percentage was 1.5.

It will be seen from Table 4.10 that, while the standard method always requires more shear reinforcement than BS 8110, the difference between the variable truss angle method and BS 8110 is largest at a shear stress of about 3 N/mm^2 where EC2 requires about 60 per cent of that required by BS 8110. At the highest level of stress, the two methods are very similar. This illustrates the difficulties of drawing any general conclusions from comparisons of EC2 and BS 8110 in this area.

4.2.9   Shear Between the Web and Flange of Flanged Beams

BS 8110 simply specifies a minimum transverse reinforcement percentage of 0.15 across flanges to resist horizontal shear (BS 8110, Table 3.27). EC2 gives a design method in clause 4.3.2.5. This method could result in significantly more reinforcement being required across the flanges than is required by BS 8110.

4.3   Torsion

Torsion is dealt with in clause 2.4 of BS 8110: Part 2 and Chapter 4.3.3 of EC2. Careful inspection shows the two methods to be basically similar, though formulated on an apparently different basis. Both codes assume that torsion is carried by a notional truss where the tension is carried by reinforcement and the compression by notional struts within the concrete. This will be seen to be analogous to shear. The difference between shear and torsion is the three-dimensional nature of the truss needed to resist torsion. There are six elements of the methods which will be considered. These are as follows:

1. Maximum permissible torsion;
2. Provision of reinforcement to resist torsion;
3. Combination of torsion and flexure;
4. Combination of torsion and shear;

5. Cases where no torsion reinforcement is required;
6. Non-rectangular sections.

These will now be considered in turn.

4.3.1   Maximum Permissible Torsion

Both EC2 and BS 8110 give rules defining the maximum torsion that can be carried
by a section, even if reinforced for torsion. This limit is required to avoid crushing
of the notional concrete compressive struts in the notional truss assumed to be
resisting the torsion. The limiting capacity in both codes is a function of a stress,
related to the compressive strength of the concrete and the section geometry. These
two elements may be considered separately.

4.3.1.1 Stress Related to Concrete Strength
In BS 8110, the stress is limited to $0.8\sqrt{f_{cu}}$, where f_{cu} should not be taken as
>40 N/mm^2. The formulation in EC2 is somewhat more complex, but can be
written as

$$1.4(0.7 - f_{ck}/200)f_{ck}/(\gamma_m(\cot\theta + \tan\theta))$$

but not less than $0.4667 f_{ck}/(\cot\theta + \tan\theta)$.

  These two formulations can be compared most easily if it is assumed that $f_{ck} =
0.8 f_{cu}$. This has been done in Fig. 4.6. Though EC2 gives the limits to $\cot\theta$ as
being $0.4 < \cot\theta < 2.5$, it seems likely that it was not intended in this instance
to consider values of $\cot\theta$ of <1, which gives the maximum capacity.

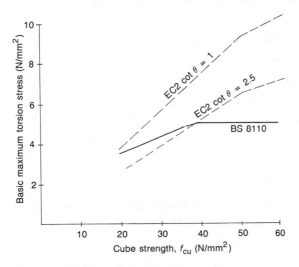

Figure 4.6 Effect of concrete strength on maximum torsion

4.3.1.2 Geometry

In BS 8110, the effect of section geometry is dealt with by multiplying the stress given above by the factor

$$h_{min}^2 (h_{max} - h_{min}/3)/2$$

where h_{min} and h_{max} are respectively the smaller and larger overall dimensions of a rectangular section.

EC2 multiplies the stress by the factor tA_k, where t is the thickness of a notional hollow section and A_k is the area enclosed within the centre-line of this notional thin-walled hollow section. The wall thickness is given by the total area of the section divided by its circumference. For a rectangular section, tA_k may be written as

$$(h_{max}h_{min})^2 (2h_{max}+h_{min})(2h_{min}+h_{max})/8(h_{max}+h_{min})^3$$

A convenient way of comparing these formulae is to non-dimensionalize these expressions by dividing them by $h_{min}h_{max}^2$ to give two 'geometry' factors, g_{EC2} and g_{8110}. These factors can be compared in Fig. 4.7. Figure 4.8 compares the combination of the stress and geometry terms for a concrete cube strength of 30 N/mm^2 and with cot θ taken as 1, which gives the maximum value in EC2. It will be seen that BS 8110 permits torsions about 1.5 times higher than EC2. This difference will decrease for cube strengths above 40 N/mm^2 due to the cut-off introduced in BS 8110. Taking cot θ as unity, the two codes will give almost identical answers for a cube strength of 60 N/mm^2.

4.3.1.3 Reinforcement for Torsion

Both codes require that reinforcement is provided in the form of closed stirrups and longitudinal bars. The formulae for this reinforcement given in the two codes are for BS 8110

$$A_{sv}/s = T/(0.8x_1 y_1 . 0.87f_{yv})$$

$$A_{s1} = A_{sv}/s(f_{yv}/f_{y1})(y_1 + x_1)$$

where T is the design ultimate torsion, A_{s1} the area of longitudinal torsion reinforcement, A_{sv} the area of *both* legs of the stirrup, x_1, y_1 the centre-line dimensions of the stirrup, s the spacing of stirrups, f_{yv} the yield strength of stirrups and f_{y1} the yield strength of longitudinal bars. For EC2

$$A_{sw}/s = T_{sd}/(2A_k f_{ywd} \cot \theta)$$

$$A_{s1} = T_{sd}u_k \cot \theta/(2A_k f_{y1d})$$

This second equation can be rearranged into a form more easily comparable to the BS 8110 equation by substituting from the first equation to give

$$A_{s1} = (A_{sw}/s)(f_{ywd}/f_{y1d})(u_k \cot \theta)$$

where T_{sd} is the design ultimate torsion, A_{sw} the area of *one* leg of a stirrup, u_k

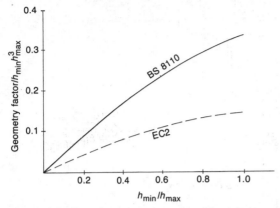

Figure 4.7 Effect of geometry on maximum torsion

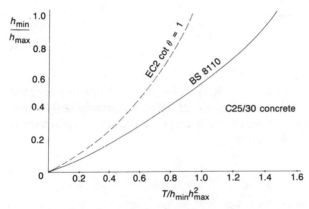

Figure 4.8 Effect of aspect ratio of section

the perimeter of centre-line of notional thin-walled section, f_{ywd} the design yield strength of stirrups and f_{yld} the design yield strength of longitudinal steel.

It will be seen that the equations for both stirrups and longitudinal bars have a very similar form. The major difference is the introduction of cot θ into the EC2 equations. One point should, however, be noted carefully. BS 8110 expresses the area of the stirrups as the area of both legs of the stirrup while EC2 uses the area of a single leg. The EC2 approach is, in fact, the more logical one since, as may be seen from Fig. 4.9, what is required is the area of reinforcement in the wall of the notional thin-walled box and this is the area of the bar forming the link. It is believed that BS 8110 uses the area of both legs so as to avoid confusion with shear design, where it is quite correct to use the area of the two legs. BS 5400 effectively uses the EC2 formulation.

The two approaches can probably best be compared by means of a simple example.

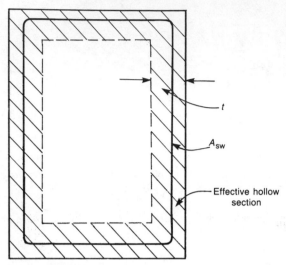

Figure 4.9 Definition of A_{sw}

Example 4.2

A rectangular section 300 mm wide by 500 mm overall depth is to be designed to withstand a design ultimate torsion of 25 kN m. Assuming 460 N/mm² reinforcement with a cover of 25 mm to the stirrups, calculate the required areas of stirrup and longitudinal reinforcement.

BS 8110

Assuming that 10 mm diameter stirrups will be used, then

$$x_1 = 300 - 2 \times 25 - 10 = 240 \text{ mm}$$

$$y_1 = 500 - 2 \times 25 - 10 = 440 \text{ mm}$$

Hence

$$A_{sv}/s = 25 \times 10^6/(0.8 \times 0.87 \times 460 \times 240 \times 440) = 0.74$$

$$A_{s1} = 0.74 \times 460/460 \times (240 + 440) = 503 \text{ mm}^2$$

EC2

The effective thickness of the notional thin-walled section, t, is given by

$$t = 300 \times 500/2(300 + 500) = 93.75 \text{ mm}$$

From this

$$A_k = (300 - 93.5)(500 - 93.5) = 83\,789 \text{ mm}^2$$

and

$$u_k = 2(300 - 93.75) + (500 - 93.75) = 1225 \text{ mm}$$

Hence, assuming cot $\theta = 1$,

$$A_{sw}/s = 25 \times 10^6/(2 \times 83\,789 \times 460/1.15 \times 1) = 0.373$$

and

$$A_{sl} = 0.373 \times 460/460 \times 1225 \times 1 = 457 \text{ mm}^2$$

Remembering that EC2 gives the area of one leg of the stirrup, the stirrup area comparable with BS 8110 is $2 \times 0.373 = 0.746$. It will be seen that the reinforcement areas predicted by the two codes are very similar.

The above example assumed that cot $\theta = 1$. To see the influence of other values of cot θ, Table 4.11 has been constructed.

Table 4.11 Effect of varying cot θ

Cot θ	$2A_{sw}/s$ (mm²/mm)	A_{sl} (mm²)
0.4	1.865	183
0.6	1.243	274
0.8	0.933	366
1.0	0.746	457
1.25	0.597	571
1.67	0.447	763
2.5	0.298	1142

It will be seen that, as cot θ increases, the amount of stirrup reinforcement decreases and the amount of longitudinal reinforcement increases. The major effect of varying the truss angle is thus to redistribute the reinforcement. It is not possible to say where the most economical arrangement lies, but it could well be in the region of cot $\theta = 1$. Certainly, from the point of view of crack control, a reasonable balance between the longitudinal and stirrup reinforcement seems likely to be beneficial.

4.3.3   Combined Torsion and Flexure

For normal members, both codes use the same approach: design is carried out separately for flexure and torsion and the resulting areas of reinforcement are added.

4.3.4   Combined Torsion and Shear

In both codes the amounts of stirrup reinforcement for torsion and shear are assessed separately and the areas summed. In EC2, the same value of cot θ should be used for both shear and torsion.

Both codes provide limits to the amount of combined shear and torsion that can be supported, but the rules are different. In BS 8110, the sum of the torsion and the shear stress is limited to $0.8\sqrt{f_{cu}}$, while in EC2 the limit is given by

$$(T_{sd}/T_{Rd1})^2 + (V_{sd}/V_{Rd2})^2 \leq 1$$

The two rules can conveniently be compared graphically and this is done in Fig. 4.10. It will be seen that BS 8110 is the more conservative. However, since T_{Rd1} in EC2 is generally lower than the equivalent factor in BS 8110, the two effects will tend to cancel out.

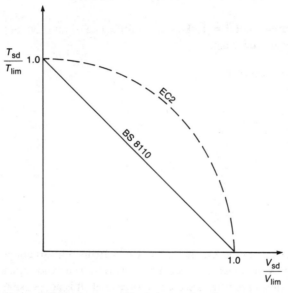

Figure 4.10 Maximum combined shear and torsion

4.3.5 Cases where no Torsion Reinforcement is Required

BS 8110 requires no reinforcement for torsion if

$$T \leq 0.067 f_{cu} h_{min}^2 (h_{max} - h_{min}/3)/2$$

The rules in EC2 are rather more complex. No reinforcement for shear and torsion is required beyond the minimum shear reinforcement if the following two conditions are met:

$$T_{sd} \leq V_{sd} b_w/4.5 \quad \text{and} \quad V_{sd}(1+(4.5T_{sd})/(V_{sd}b_w)) \leq V_{Rd1}$$

Figure 4.11 provides a graphical interpretation of the EC2 rule.

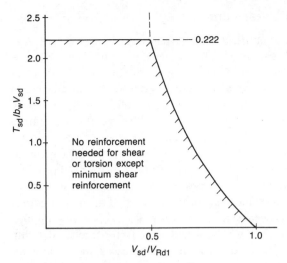

Figure 4.11 EC2 rule for no requirement for torsion reinforcement

4.3.6   Non-rectangular Sections

The two codes may be assumed to be the same for the treatment of sections which can be made up of a series of rectangles, such as T or L sections. BS 8110 does not cover other forms of section.

4.4   Punching Shear

Punching shear is the local shear failure round a column in a flat slab or round a concentrated load. Like normal shear in beams, the design methods for punching shear are basically empirical in both BS 8110 and EC2. Both methods have been justified by comparison of the predictions with very large quantities of test data. It is therefore, at first sight, surprising that the two codes contrive to give rather different answers; especially since the basic approaches in the two codes are very similar.

In both codes, the design procedure consists of the following steps:

1. Establish the design effective shear force.
2. Establish the first or critical perimeter.
3. Calculate the shear capacity of the critical perimeter.
4. If the shear capacity of the critical perimeter is less than the design effective shear then calculate the required quantity of shear reinforcement. Check subsequent perimeters in the same way until a perimeter is reached which does not require shear reinforcement.
5. Check that the shear force does not exceed the maximum permissible shear.

Each of these steps will be considered in turn and the differences and similarities between the codes examined.

4.4.1   Design Effective Shear Force

The basic shear force will be established from the analysis of the structure. Any differences in the values will simply arise from the slightly different values of the partial safety factors on the loads and the slightly more onerous requirement in EC2 that the 'adjacent spans loaded' and 'alternate spans loaded' arrangements of loads be considered whereas, in BS 8110, it will generally only be necessary to consider the single load case of maximum load on all spans. It is recognized that where a connection between a slab and a column imposes a moment transfer between the slab and the column then the punching shear capacity will be reduced. Both codes deal with this by increasing the design shear force. BS 8110 provides formulae for the prediction of this increase as a function of the moment transfer and also 'deemed to satisfy' rules which may be used instead of the more rigorous calculation. EC2 only provides 'deemed to satisfy' rules. The two sets of simplified rules are compared in Table 4.12. It will be seen that the factors are similar with the EC2 values being slightly simpler to apply but, in some cases, more conservative. The effective shear is obtained by multiplying the basic shear by the factors given in Table 4.12.

Table 4.12   Factors for moment transfer

Structural situation	BS 8110	EC2
Internal column	1.15	1.15
Edge column (a) bending parallel to edge	1.4	1.4
(b) bending perpendicular to edge	1.25	1.4
Corner columns	1.25	1.5

4.4.2   Definition of Perimeters

Both codes locate the critical perimeter at a distance of $1.5d$ from the face of the loaded area or column. BS 8110 uses a rectangular perimeter while EC2 uses a perimeter with rounded corners (see Fig. 4.12). The BS 8110 perimeter was chosen for simplicity and because it was felt that shear reinforcement, if needed, would practically have to be provided on rectangular perimeters. The EC2 perimeter reflects the actual form of the likely failure surface more correctly. The EC2 perimeter is $2.58d$ shorter than the BS 8110 perimeter. As a percentage, this difference is usually fairly small, probably generally < 15.

4.4.3   Calculation of Shear Capacity of Perimeter

Both codes calculate the capacity of a perimeter without shear reinforcement using the same equations as they use for beams. These equations are as follows. For BS 8110:

$$v_c = 0.79(\rho f_{cu}/25)^{1/3}(400/d)^{1/4}/\gamma_m$$

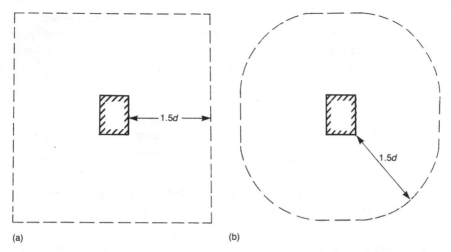

Figure 4.12 Punching shear perimeters. (a) BS 8110; (b) EC2

In this equation, γ_m should be taken as 1.25, ρ should not be taken as >3, f_{cu} should not be taken as $>40 \, \text{N/mm}^2$ and $400/d$ should not be taken as <1.0. For EC2:

$$v_{Rd1} = 0.0452 f_{cu}^{2/3}(1.6-d)(1.2+0.4\rho)/\gamma_m$$

In order to make comparisons easy, it has been assumed that f_{ck} is equal to $0.8 f_{cu}$ and the equation adjusted to be in terms of f_{cu}. In the equation, d is in metres and should not be taken as $>0.6 \, \text{m}$ and ρ should not be taken as >2 per cent. Unlike BS 8110, which simply requires that an average value be taken for ρ, EC2 calculates an effective steel percentage from the relation

$$\rho_{eff} = \sqrt{\rho_x \rho_y}$$

Here γ_m is taken as 1.5 for shear in EC2 rather than the value of 1.25 used in BS 8110. The earlier section on shear in beams (section 4.2) compares these formulae.

Both codes calculate the shear capacity of the perimeter as

$$V_{sd} = v_c U d$$

where U is the length of the perimeter.

4.4.4   Provision of Shear Reinforcement

Both codes calculate the required amount of shear reinforcement by identical formulae, but the detailing rules differ. EC2 requires the first perimeter of links to be at $0.5d$ from the face of the column, while BS 8110 used to ask for $0.75d$. This, however, has been changed by a recent amendment to BS 8110 which brings it into line with EC2 in this respect. The spacing of subsequent perimeters of links is given as $0.75d$ in BS 8110. In EC2 there is an element of obscurity. Figure 5.17a in EC2 indicates a spacing of $0.75d$; however, clause 5.4.3.3(4) states that the maximum longitudinal spacing should not exceed the following:

$$0.8d \quad \text{where } v_{sd} < 0.2v_{Rd2}$$
$$0.6d \quad \text{where } 0.2v_{Rd2} < v_{sd} < 0.67v_{Rd2}$$
$$0.3d \quad \text{where } v_{sd} < 0.67v_{Rd2}$$

The value of v_{Rd2} will be considered below, but the limits of $0.6d$ and $0.3d$ are frequently likely to govern the spacing.

Both codes state that the calculated area of reinforcement should be provided within the perimeter considered.

4.4.5   Maximum Shear

BS 8110 currently limits the shear that can be supported in punching to the shear force that corresponds to a stress of $0.8\sqrt{f_{cu}}$ on the perimeter of the column. This is, however, likely to change as experimental work has shown it to be unconservative. EC2 limits the shear to that corresponding to a stress of $1.6v_{Rd1}$ on the critical perimeter. This is a much more severe restriction than the current BS 8110 limit. It should be noted, however, that revisions are being considered to BS 8110 which would make the codes more nearly comparable in this respect.

There are a number of other differences between the provisions of the two codes which should be noted. The two most important are as follows:

1.   The treatment of edge columns in BS 8110 where the moment that can be transferred between slab and column is limited. This has no parallel in EC2.
2.   The requirement in EC2 that reinforcement over the columns shall be able to support minimum moments which are a function of the applied shears. Though not stated, it seems likely that this is required to ensure that a local 'fan' type of flexural yield line failure cannot occur prematurely. Calculation suggests that the rules given in BS 8110 for the arrangement of reinforcement in the column strips serves the same purpose.

4.4.6   Comparison of Punching Shear Capacities of EC2 and BS 8110

It is not easy to produce a clear general comparison of the practical consequences of the two codes because of the number of variables involved. The following comparison has been taken from the calibration studies carried out by the BCA for BRE and compares the provisions of the two codes for internal columns in slabs with equal spans in both directions.

The total moment over a support can be written as

$$M_{tot} = k_m l^3 n_u$$

where k_m is a bending moment coefficient and n_u the design ultimate load per unit area. Similarly, the design shear force can be written as

$$V = k_v l^2 n_u$$

where k_v is a shear force coefficient.

The total area of top reinforcement can be calculated approximately as

$$A_s = M_{tot}/(0.9d \times 0.87 f_y) = k_m lV/(0.78 d k_v f_y)$$

BS 8110 assumes that 75 per cent of the reinforcement is provided in the column strip with two-thirds of this provided within the central half of the strip. This means that 50 per cent of the total reinforcement is provided within a strip of width equal to 25 per cent of the span. Substituting for this gives the reinforcement ratio in the region of the column as

$$\rho = 2 k_m V/0.78 k_v f_y d^2$$

If x is the distance from the centre of the column to the perimeter then the length of the perimeter according to BS 8110 is $8x$ and hence

$$V_{eff} = 8dx \times 0.632 (\rho f_{cu}/25)^{0.33} (400/d)^{0.25}$$

Assuming $f_{cu} = 30$, $f_y = 460$, $d = 200$ in assessing the depth correction $(400/d)^{0.25}$ and since $V_{eff} = 1.15V$ for internal columns, this can be simplified to

$$V^2/d^4 = 94.7(x/d)^3 k_m/k_v$$

From BS 8110, reasonable values for k_m and k_v are 0.086 and 1.05 respectively, giving

$$V^2/d^4 = 7.75(x/d)^3$$

For simplicity, it will be assumed that the steel percentage formula will be the same for EC2 as in BS 8110 and that the perimeter is 15 per cent smaller. On this basis, and using the same assumptions as before for material strengths, etc. we can write

$$V_{eff} = 2.849(1.2 + 40\rho)dx$$

Substituting for the reinforcement ratio gives

$$V/d^2 = 2.477(1.2 + 0.223 k_m V/k_v d^2)x/d$$

Owing to the different load arrangements used in EC2, comparable values for k_m and k_v for EC2 are 0.104 and 1.1. Substituting for these gives

$$V/d^2 = 2.98(1 + 0.0175 V/d^2)x/d$$

These two equations can now be used to compare the two codes. It will be seen from Fig. 4.13 that EC2 is very substantially more conservative than BS 8110

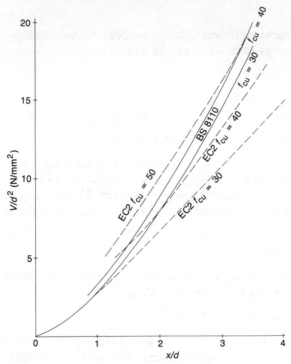

Figure 4.13 Comparison of punching shear capacity of slabs without shear reinforcement. $x =$ Distance to perimeter from centre of loaded area at which $v = v_c$ or $V_{sd} = V_{Rd}$

with $f_{cu} = 30 \, \text{N/mm}^2$. It will be seen that at $40 \, \text{N/mm}^2$ the two codes are fairly similar and that at $50 \, \text{N/mm}^2$ BS 8110 is more conservative than EC2. This final result is partly due to the greater influence which EC2 gives to concrete strength and partly to the cut-off in BS 8110 at a concrete strength of $40 \, \text{N/mm}^2$.

4.4.7 Punching Shear Example

The first interior support of a flat slab with four 6 m spans in both directions will be designed using the provisions of BS 8110 and EC2. The characteristic imposed and permanent loads on the slab are, respectively, $5 \, \text{kN/m}^2$ and $5 \, \text{kN/m}^2$, based on a 200 mm thick slab. The columns are 350 mm square. C30/37 concrete and 460 grade reinforcement have been chosen.

4.4.7.1 Design According to BS 8110

An analysis was carried out using the equivalent frame method. For this structure the simplified load arrangement for slabs may be used, and thus analysis only has to be carried out for the single load case of maximum load on all spans. The resulting bending moment diagram is then redistributed downwards by 20 per cent. This analysis led to the following design actions at the first interior support.

Total design moment in panel = 276.68 kN m
Design shear = 616.46 kN

Assuming a nominal cover of 20 mm and that 16 mm bars will be used gives effective depths in the two directions of 172 and 156 mm. This gives an average effective depth of 164 mm.

Seventy-five per cent of the total moment is taken into the column strip, giving a moment of 204.51 kN m. Using the design charts in BS 8110 gives areas of reinforcement in the two directions of 3302 mm^2 and 3650 mm^2. Two-thirds of these areas should be placed within a band centred on the column of width equal to half the column strip. The resulting areas of reinforcement required within a 1.5 m wide strip are respectively 2201 mm^2 and 2434 mm^2. These areas will be met by using 12 No.16 mm bars at 135 mm centres both ways.

The average reinforcement percentage in the region of the column is thus

$$201 \times 100/(135 \times 164) = 0.908 \text{ per cent}$$

This corresponds to a value for v_c of 0.872 N/mm^2.

The length of the critical perimeter is given by

$$u = 4(350+3 \times 164) = 3368 \text{ mm}$$

For internal columns the multiplier to allow for moment transfer is 1.15 and hence the design shear stress on the critical perimeter is

$$v = (1.15 \times 616.46 \times 1000)/(3368 \times 164) = 1.283 \text{ N/mm}^2$$

The shear on the perimeter of the column is

$$v_{max} = (1.15 \times 616.46 \times 1000)/(164 \times 4 \times 350) = 3.088 \text{ N/mm}^2$$

This is less than $0.8\sqrt{f_{cu}} = 4.87$ N/mm^2 so design is possible.

Since v exceeds v_c, shear reinforcement will be required. Since v is less than $1.6v_c$, the total area needed to reinforce for failure at the first perimeter is given by

$$3368 \times 164 \times (1.283 - 0.872)/(0.87 \times 460) = 568 \text{ mm}^2$$

This should be provided on two perimeters and the links should not be spaced round the perimeters at spacings $> 1.5d = 246$ mm.

It has been decided to use 6 mm bars at 225 mm centres round both perimeters. This requires 20 bars, giving an area of 466 mm^2.

Shear now needs to be checked at successive perimeters at $0.75d$ intervals until v is found to be $< v_c$. It will be found that one further perimeter will need reinforcing. The spacing limit will require that, again, the reinforcement used will be 6 mm bars at 225 mm centres.

Figure 4.14a shows the resulting shear reinforcement.

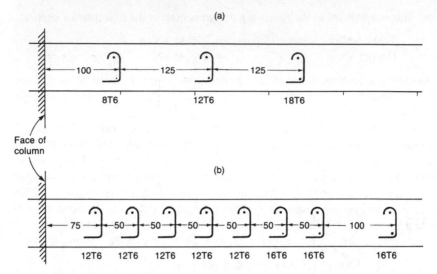

Figure 4.14 Comparison of shear designs from punching shear example. (a) Design to BS 8110;
(b) design to EC2

4.4.7.2 Design according to Eurocode 2

EC2 provides no guidance on how one should analyse flat slabs and arrange the reinforcement, so it has been assumed that the BS 8110 equivalent frame method may be used. EC2, however, requires analyses to be done for alternate spans loaded and adjacent spans loaded. The resulting analysis gives the following design actions:

$$\text{Total design moment across panel at support} = 329.88 \text{ kN m}$$
$$\text{Design shear force} = 602.5 \text{ kN}$$

As with the BS 8110 design, it has been assumed that 75 per cent of the moment is in the column strip and that two-thirds of the resulting area of reinforcement is placed in the central half of the strip. This leads to steel areas in the 1.5 m strip over the column of 2683 and 2933 mm^2 in the directions having the larger and smaller effective depths respectively. These areas will be met by providing 14 No.16 mm bars in each direction, giving reinforcement ratios respectively of 0.0109 and 0.0120. The effective reinforcement ratio is then calculated as the square root of the product of the ratios in the two directions. This gives 0.0114.

EC2 requires that a strip 0.3 × the panel width wide over the column should be reinforced to support a moment of at least $0.125V_{sd}$ kN m/m. This gives a moment of 75 kN m/m which is considerably below the reinforcement required by the analysis.

The capacity of the section without shear reinforcement is now given by

$$v_{Rd1} = 0.34 \times (1.6 - 0.164) \times (1.2 + 40 \times 0.0114) = 0.808 \text{ N/mm}^2$$

The length of the critical perimeter is given by

$$u = 4 \times 350 + \pi \times 3 \times 164 = 2946 \, \text{mm}$$

The multiplier for moment transfer is 1.15 for internal columns, as in BS 8110. The design shear stress on the perimeter is thus:

$$v_{sd} = (1.15 \times 602.5 \times 1000)/(2946 \times 164) = 1.434 \, \text{N/mm}^2$$

The maximum shear allowable, according to EC2 as drafted is $1.6v_{Rd1} = 1.292 \, \text{N/mm}^2$. This is less than the design shear stress; however, the NAD amends this provision to conform to the recent amendment to BS 8110. This increases the maximum to $2v_{Rd1} = 1.616 \, \text{N/mm}^2$. This exceeds v so the design is possible.

Since v_{sd} exceeds v_{Rd1}, shear reinforcement is required. The total area of shear reinforcement for the first perimeter according to EC2 as drafted is given by

$$A_{sw} = (1.434 - 0.808) \times 2946 \times 164/(0.87 \times 460) = 756 \, \text{mm}^2$$

This will be provided by 28 No.6 mm bars. It will be found that reinforcement is required on two further perimeters.

The NAD, however, has amended the equations for the calculation of shear reinforcement. The equation used above holds where v_{sd} is $< 1.6v_{Rd1}$, but for higher shears the following formula is given:

$$A_{sw} = (v_{sd} - 1.4v_{Rd1})ud/(0.3 \times 0.87f_{yk})$$

Since v_{sd} exceeds $1.6v_{Rd1}$ it is this revised equation that applies and the shear reinforcement area is given by

$$A_{sw} = (1.434 - 1.4 \times 0.808) \times 2946 \times 164/(0.3 \times 0.87 \times 460) = 1218 \, \text{mm}^2$$

This amendment to EC2 in the NAD seems somewhat perverse since it has led, in this case, to EC2 requiring more than double the reinforcement required by BS 8110.

While EC2 has no requirements for the spacing of the shear reinforcement around the perimeter, it does have rules for the spacing of the bars perpendicular to the perimeter. This gives the maximum spacing as a function of v_{sd}/v_{Rd2} where v_{Rd2} is equal to $2v_{Rd1}$. For shear on the critical perimeter, this ratio is equal to 0.88 and, in this case, the spacing of the shear reinforcement should not exceed $0.3d$. This is approximately 50 mm. Since the first perimeter of shear reinforcement must be not more than $d/2$ from the column face, it will be necessary to use four perimeters of shear reinforcement to reinforce for this first critical perimeter. The most convenient arrangement of reinforcement seems to be to use 12 No.6 mm bars on each of the four perimeters.

Further checks need to be carried out for perimeters at $0.75d$ intervals beyond the first perimeter until a perimeter is reached which does not require reinforcement. It is found that two further perimeters require reinforcing. Owing to the spacing rules, this requires another four perimeters of reinforcement. Though these perimeters could theoretically contain less than 12 bars, the spacings become ridiculously large. A practical limit on spacing of 300 mm has been adopted, leading to the use of 16 bars in the three outermost perimeters.

The details of the shear reinforcement are shown in Fig. 4.14b.

The above example shows that the basic approach of the two codes is very similar but that, as suggested by the parameter study, EC2 is significantly more conservative. This conservatism is markedly increased by the amendment made in the NAD and by the spacing requirements. It is hard to feel that either the amendment or the very close spacing of the shear reinforcement are really necessary.

4.5   Slender Columns

When a column is loaded it will deflect. This deflection increases the eccentricity of the load and hence the moment. The magnitude of this deflection will depend upon the end conditions of the column, the slenderness and the initial applied moments. In short, or stocky, members the deflection is so small that it makes no significant difference to the strength, while with increasing slenderness the effect becomes increasingly important. Design methods for these effects, commonly called 'second-order effects', generally consist of two basic elements:

1.   Establishing whether or not the deflections are likely to be significant;
2.   If they are significant, provisions to take account of them.

Both BS 8110 and EC2 include these two stages and, for most common types of member, employ very similar methods to do so. The provisions of BS 8110 are undoubtedly presented with much greater brevity and clarity though, in a number of areas, EC2 is philosophically more rigorous. It should be noted, however, that EC2 does not give specific design methods for designing unbraced sway structures whereas BS 8110 does. One very useful feature of EC2 is a set of flow charts included in Appendix 3. Without these the intentions of the provisions would be very much more difficult to discern.

Elements (1) and (2) above will now be considered in turn and the requirements of the two codes compared.

4.5.1   Is the Structure Slender?

A structure is deemed to be slender if the second-order effects are significant. EC2 provides a definition of what is meant by significant in clause 4.3.5.1(5) where it states that '. . . second-order effects should be considered if the increase above the first-order bending moments due to deflections exceeds 10%'. While this statement is of rather academic interest, it is nevertheless useful to have this point clarified.

The first step in deciding on whether or not the structure is slender is to classify the structure, as this defines the nature of the deflection to be expected. In EC2 structures are classified as 'sway' or 'non-sway' and as 'braced' or 'unbraced'. In theory, at least, this system admits four possible classes: braced—sway, braced—non-sway, unbraced—sway and unbraced—non-sway. BS 8110 takes a simpler approach and implicitly assumes that a braced structure will be non-sway and an unbraced structure will be a sway structure. While the first of these assumptions will be true in all normal circumstances, the second may not be and to assume that all unbraced frames are sway frames could, on occasion, be unnecessarily conservative.

A sway structure is one where the structure may sway significantly so that the tops of the columns deflect relative to the bottoms. In a non-sway structure it may be assumed that the tops of the columns do not deflect relative to the bottoms. A braced structure or element is one where any horizontal loads can be assumed to be carried by other 'bracing' elements. These definitions are set out explicitly in EC2, but are implicit in BS 8110 also. In BS 8110 the decision on classification is taken by inspection. This can generally also be done in EC2, but methods are also given whereby the classification can be checked by calculation. Thus, EC2 states that, for a structure to be considered as braced, the bracing elements should be sufficiently stiff to attract at least 90 per cent of any horizontal forces applied to the structure. Similarly, structures should be considered as sway structures if the displacements of the tops of the columns increases the moments at the tops by more than 10 per cent. Formulae are given in Appendix 3 to help with this classification. Despite these extra definitions and formulae, it is thought that it will, in most cases, be possible to classify structures by inspection as is done in BS 8110.

If a structure has been classified as 'sway' then EC2, unfortunately, provides little more help, it being implied that 'rigorous' methods of analysis will be necessary. The bases for such analyses are set out in other parts of the code but, since fairly complex computer programs would be needed to carry them out, they will not be considered further here. The rest of this section will, therefore, be concerned with the design of braced, non-sway columns.

Having established that the column being considered is, indeed, a braced, non-sway column, the decision on whether or not it is a slender column is taken on the basis of its slenderness ratio. The slenderness ratio is, however, defined differently in the two codes. In BS 8110 it is defined as the ratio of the effective height to the smallest lateral dimension, while in EC2 it is the ratio of the effective height to the radius of gyration. For a rectangular column, the slenderness ratio by the EC2 definition is $\sqrt{12}$ ($= 3.46$) times the slenderness ratio according to BS 8110. The EC2 definition is considered to be more rigorously correct than the BS 8110 definition and it allows a more logical estimation of the slenderness of non-rectangular columns. The concept of effective length is the same in both codes, but EC2 does not include the simplified rules for its assessment as are given in Tables 3.21 and 3.22 of BS 8110. Instead, it provides a nomogram which provides solutions to the more rigorous equations given in Part 2 of BS 8110.

In EC2, the actions to be taken, depending on the slenderness ratio, are as follows:

1. If the slenderness ratio <25. The column is not slender and no action is required to deal with second-order effects.
2. If $25 <$ slenderness ratio $< 25(2 - e_{01}/e_{02})$. The column ends should be designed to withstand a moment of at least $N_{sd}h/20$.
3. If $25(2 - e_{01}/e_{02}) <$ slenderness ratio < 140. In this case, specific measures should be taken to deal with second-order effects. These will be considered in section 4.5.2.

In the above, e_{01} is the numerically smaller end eccentricity, e_{02} the numerically larger end eccentricity, N_{sd} the design axial load and h the overall dimension of the column in the direction considered.

In (3) above, since the eccentricities at top and bottom are likely to be in opposite directions, one will be positive and one negative so that the limiting slenderness ratio given by the expression will generally be in the range $50-75$. Using the BS 8110 definition of slenderness ratio, this corresponds to a range of slenderness of between 14.4 and 21.7. Since BS 8110 requires design for at least a minimum moment of $Nh/20$ regardless of slenderness and requires second-order effects to be taken into account when the slenderness ratio exceeds 15, it will be seen that BS 8110 and EC2 will generally produce the same result for slenderness ratios of up to 15 (BS 8110 definition).

In summary, in assessing whether or not a column is slender, both EC2 and BS 8110 will come to very similar conclusions.

4.5.2   Design for Second-order Effects

The method likely to be used by most designers when using EC2, called the 'model column method' and the additional moment method used in BS 8110 are basically the same. Both methods aim to predict the curvature at the critical section in the column at failure and hence, by assuming a sinusoidal deflected shape, to predict the ultimate deflection. This deflection, when multiplied by the design load, gives the design additional or second-order moment. This has to be added to the first-order moment to give the design moment. Since the maximum second-order moment will occur around mid-height of the column, an estimate of the first-order moment at this level is required. Both codes assume that this moment is given by

$$M_i = 0.6M_{02} + 0.4M_{01} \text{ but not less than } 0.4M_{02}$$

where M_{01} and M_{02} are respectively the numerically smaller and greater end moments on the column. EC2 expresses this relationship in terms of eccentricities rather than moments, but is otherwise identical.

The ultimate deflection is obtained from the following expressions in the two codes.

$$\text{BS 8110   ult. deflection} = kh(l_0/h)^2/2000$$

EC2   ult. deflection $= 2k_1k_2f_{yd}d(l_0/d)^2/9E_s$

In the EC2 equation k_2 is identical to k in the BS 8110 equation and is given (in BS 8110 notation) by the equation

$$k = (N_{uz}-N)/(N_{uz}-N_{bal}) < 1.0$$

where N_{uz} is the axial load capacity of the column in the absence of any moment and N_{bal} the load corresponding to a balanced condition.

In EC2 k_1 is a factor which takes a value of 1.0 for situations where the slenderness ratio (EC2 definition) is greater than 35. It will therefore be 1.0 for all practical situations. It can only take a value other than 1.0 in columns where the eccentricities at top and bottom have the same sign. Assuming 460 N/mm^2 grade reinforcement, that d is $0.9h$ and that the modulus of elasticity of the reinforcement is 200 000 N/mm^2 and substituting these values into the EC2 equation gives

Ult. deflection $= kh(l_0/h)^2/1761$

This will be seen to be about 13 per cent greater than the BS 8110 value. EC2 requires that an additional 'accidental' eccentricity is also included. This takes a value of

$$e_a = \alpha_n v l_0/2$$

where v is the greater of $1/(100h_t)$ or $1/200$, h_t the overall height of the structure, α_n is $(1+1/n)/2$ and n is the number of vertically continuous elements which act together (i.e. generally the number of columns and walls in the structure).

In the simplest possible case of a square column subjected to uniaxial bending, the two codes thus require design for the following loads and moments:

EC2   Design axial load $= N_{sd}$
           Design moment $= N_{sd}(e_0+e_a+e_2)$
BS 8110   Design axial load $= N$ (equivalent to N_{sd})
           Design moment $= M_i+M_{add}$ (equivalent to $N_{sd}(e_0+e_2)$)

These are very similar with EC2 being likely to be slightly more conservative than BS 8110.

Buckling being a somewhat random process, it is possible for deflections to occur about either axis where the column is square and the axial load small. For rectangular sections, there is a high probability of deflection about the minor axis even when the bending is about the major axis. These possibilities have to be considered in design and the procedures set out in the two codes differ significantly.

In BS 8110 the procedure may be summarized as follows.

1. If a column is bent only about its minor axis, there is no necessity to consider either bending or deflection about the major axis.
2. If a column is bent about its major axis then:
   (a) If $(l_e/b) < 20$ then design is carried out for uniaxial bending about the

major axis, but the second-order eccentricity is calculated on the basis of the smaller section dimension.

(b) If $(l_e/b) > 20$ then design has to be carried out for biaxial bending. The moment about the major axis is the initial (first-order) moment plus the second-order moment calculated on the basis of the larger section dimension. The moment about the minor axis is the second-order moment calculated on the basis of the smaller section dimension. The column is thus assumed to be deflecting in both directions at once.

3.   If the column is bent about both axes then the moment about each axis is taken as the initial moment plus the second-order moment based in the appropriate section dimensions.

For EC2 the procedures may be summarized as follows:

1.   Any column subjected to uniaxial bending should be checked for the possibility of it deflecting about the other axis. This is done by considering the possibility of deflections occurring about each axis in turn. Thus, two load cases need to be considered:

(a) An eccentricity of $e_0 + e_a + e_2$ about the axis of principal bending and zero about the other axis.

(b) An eccentricity of $e_a + e_2$ about the other axis and e_0 about the principal axis of bending.

In (a) and (b) above, e_2 is calculated on the basis of the section dimension perpendicular to the axis of bending considered.

2.   For columns subjected to biaxial bending, a similar principle is applied and the two load cases to be considered are:

(a) An eccentricity of $e_{0x} + e_a + e_{2x}$ about the x-axis and an eccentricity of e_{0y} about the y-axis.

(b) An eccentricity of e_{0x} about the x-axis and an eccentricity of $e_{0y} + e_a + e_{2y}$ about the y-axis.

The EC2 approach is generally less conservative than that of BS 8110 since the additional moments are considered to act only about one axis at a time. A serious disadvantage with the EC2 method is that EC2 gives no method for the design of sections subjected to biaxial bending other than applications of the basic assumptions for flexural behaviour. To overcome this for some situations, the following simplification is given. If

$$(e_x/h)/(e_y/b) < 0.2 \quad \text{or} \quad (e_y/b)/(e_x/h) < 0.2$$

then separate uniaxial checks can be carried out for bending about the two axes, ignoring any moment about the other axis. There is, however, a further limitation that, if the eccentricity about the major axis exceeds $0.2h$, then a fictitious reduced value should be used for h when checking for bending about the minor axis.

An extensive parameter study was carried out for the BRE to compare the two codes over the practical range of columns. This showed that, for the most common types of column, the provisions of the two codes led to very similar column designs. This can be seen for a typical case in Fig. 4.15. Figure 4.16 shows the whole range of results obtained in the study and it will be seen that, while there can be significant differences, in the great majority of cases there are not. The significant differences tend to occur due to the different treatment of rectangular and biaxially bent columns.

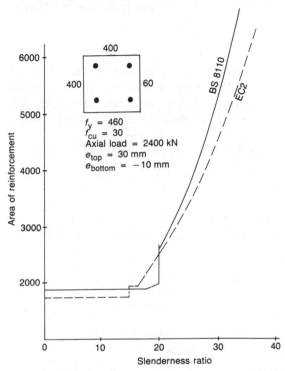

Figure 4.15 Typical comparison of buckling provisions of EC2 and BS 8110

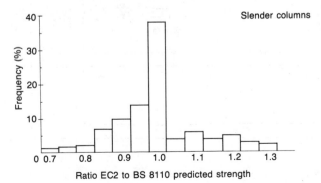

Figure 4.16 Comparisons of EC2 and BS 8110 from parameter study

Example 4.3

The design is going to be considered of the 9 m high columns in the braced frame shown in Fig. 4.17. All beams and columns are rectangular with section dimensions of 500×300 mm. Analysis of the structure has given moments about the major axis at the top of the column of 140 kN m and at the bottom -80 kN m. The first-order bending moments about the minor axis are zero. The cube strength of the concrete is 45 N/mm^2 and the steel may be assumed to be 460 N/mm^2.

Design to BS 8110

The first step is to establish the effective length of the column. Since the beams attached to the column have the same depth as the column then the end conditions may be taken as condition 1 at both top and bottom. Hence, from Table 3.21 in BS 8110, the factor β is 0.75. This gives the effective length as $0.75 \times 9000 = 6750$ mm.

The slenderness ratios in the two directions may now be calculated. They are as follows:

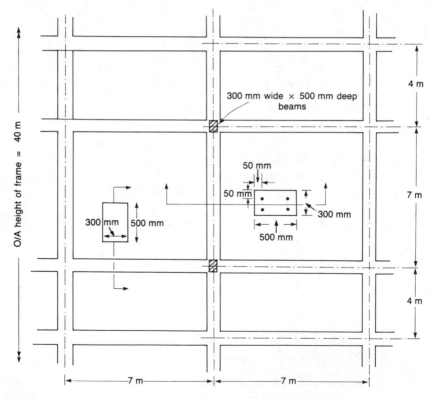

Figure 4.17 Frame used in example

For major axis bending: $l_e/h = 6750/500 = 13.5$
For minor axis bending: $l_e/b = 6750/300 = 22.5$

Since one of these values exceeds 15, the structure is slender (clause 3.8.1.3).

Since bending is about the major axis and l_e/h is less than 20, it is only necessary to design for uniaxial bending about the major axis (clause 33.8.3.3). The ultimate deflection is taken as

$$a_u = (l_e/b)^2 hk/2000 = 22.5^2 \times 500/2000 \times k = 126.6k$$

The initial moment is the greater of $0.4M_1 + 0.6M_2$ or $0.4M_2$ and is

$$0.41 \times -80 + 0.6 \times 140 \quad \text{or} \quad 0.4 \times 140 = 52 \quad \text{or} \quad 56$$

Hence the initial moment is 56 kN m.

The design moment is now the sum of the initial and additional moments which is given by

$$M = 56 + 4000 \times 126.6/1000k = 56 + 506k \text{ kN m}$$

The value of k and hence the design moment and the area of reinforcement can be found iteratively using Design Chart 44 in Part 3 of BS 8110. The procedure is as follows: Assume $k = 1$. This gives a moment of $56 + 506 = 562$ kN m.

$$M/bh^2 = 7.49 \quad \text{and} \quad N/bh = 26.67$$

From the chart

$$k = 0.55 \text{ hence } M = 334 \text{ and } M/bh^2 = 4.45$$

From the chart

$$k = 0.4, \text{ etc.}$$

The value of k finally settles at 0.3, giving $M = 207.8$ and $M/bh^2 = 2.8$. This corresponds to a reinforcement percentage of 3.1, giving a steel area of 4650 mm^2.

Design to EC2

Calculation of the effective length is more complicated using EC2 as it is necessary to use the nomogram in clause 4.3.5.3.5. Since the elastic modulus of the concrete may be assumed to be the same in all members and since all members have the same section dimensions, $k_a = k_b$ and is given, using Eq. 4.60, as

$$k = (\tfrac{1}{9} + \tfrac{1}{4})/(\tfrac{1}{7} + \tfrac{1}{7}) = 1.26$$

From the nomogram the coefficient β can be found to be 0.8, and hence the effective length of the column for bending about the major axis is $0.8 \times 9000 = 7200$ mm. An effective length should also be calculated for bending about the minor axis. Assuming the beams connected to the tops and bottoms of the columns are 500 mm deep by 300 mm wide and are also 7 m span, then the moment of inertia of the beams will be $(500/300)^3 \times 300/500 = 2.78$ times that of the column. This gives k as

$$k = (\tfrac{1}{9} + \tfrac{1}{4})/2.78/(\tfrac{1}{7} + \tfrac{1}{7}) = 0.45$$

From the nomogram, this gives β as 0.67 and hence the effective length for minor axis bending as 6000 mm. The slenderness ratios are now as follows:

For major axis bending $7200 \times 3.46/500 = 49.8$
For minor axis bending $6000 \times 3.46/300 \ = 69.2$

It is now necessary to find the critical slenderness ratio from Eq. 4.62. The end eccentricities are obtained by dividing the end moments by the axial load. This gives $e_{01} = -20$ and $e_{02} = +35$, hence the critical slenderness ratio is given by

$$\lambda_{crit} = 25(2 + 20/35) = 64.3$$

The column is thus slender for bending about the minor axis and allowance has to be made for the second-order eccentricity.

As in the BS 8110 calculation the effective first-order moment near mid-height of the column is 56 kN m. Since the column is not slender for major axis bending, no second-order moments need be considered about this axis. Second-order moments will, however, need to be considered about the minor axis together with the accidental eccentricity, e_a.

Using Eqs 4.61 and 2.10, the accidental eccentricity can be found to be given by

$$e_a = 6000/200/2 = 15 \text{ mm}$$

This could be further reduced to allow for the number of columns acting together, but this will not be done.

The second-order eccentricity can now be calculated as

$$e_2 = 2 \times 0.87 \times 460 \times 6000^2 \times k/(0.9 \times (300 - 50) \times 200\ 000 \times 10) = 64k \text{ mm}$$

The minor axis design moment is thus

$$M_{sdy} = 4000 \times (15 + 64k)/1000 \text{ kN m} = 60 + 256k \text{ kN m}$$

Three possible loading cases should be considered:

1.   140 kN m about the major axis at the top;
2.   56 kN m about the major axis at about mid-height;
3. $60 + 256k$ kN m about the minor axis at about mid-height.

By inspection, case (3) is bound to be the critical case. The estimation of the required reinforcement area is carried out by exactly the same procedure as was used in the BS 8110 example and, in this case, gives a minor axis design moment of 145 kN m and a reinforcement ratio of 3.8 per cent, giving an area of 5250 mm^2. In this case it will be seen that EC2 gives a slightly higher steel area than does BS 8110, but the difference is only 13 per cent.

5 Serviceability Limit States

5.1 General

EC2 covers three aspects of serviceability. These are as follows:

1. Checks on the tensile steel and compressive concrete stresses;
2. Control of cracking;
3. Control of deflections.

Sections 5.2, 5.3 and 5.4 will deal with each of these in turn and compare the EC2 rules with those in BS 8110. Before doing so, however, it is necessary to make some general statements about the treatment of serviceability in the two codes.

EC2 makes it clear in each section that design for serviceability must be considered in relation to the particular nature and function of the structure being considered. No hard and fast criteria for defining the limits to satisfactory behaviour can or should be set down in a code. This approach may be seen from the following quotations from the code.

> If the proper functioning of a member is likely to be adversely affected by these [high compressive stresses], measures shall be taken to limit the stresses to an appropriate level (clause 4.4.1.1 P(1)).

> Appropriate limits, taking account of the proposed function and nature of the structure and the cost of limiting, should be agreed with the client (clause 4.4.2.1 P(5)).

> Appropriate limiting values of deflection taking into account the nature of the structure, of the finishes, partitions and fixings and upon the function of the structure should be agreed with the client (clause 4.4.3.1 P(2)).

While a similar intent can be read into some of the serviceability provisions in BS 8110, there are no such clear statements of the requirement for serviceability design to take account of the particular conditions relating to the structure in question. Certainly, the criteria set out in BS 8110 tend to be obeyed unquestioningly by very many designers.

A second difference in principle is in the definition of the appropriate loading for use in checking serviceability limits. BS 8110 is clear that the loads to be considered for all serviceability checks are the characteristic loads. EC2 defines three possible levels of loading which may be used, depending on the nature of the particular check being carried out. These are as follows.

The rare combination of loads, defined by the relation

$$G_k + Q_{k,1} + \Sigma\psi_{1,i}Q_{k,i}$$

This will normally be the same as the characteristic loads used by BS 8110.

The frequent combination

$$G_k + \psi_{1,1}Q_{k,1} + \Sigma\psi_{2,i}Q_{k,i}$$

The quasi-permanent combination

$$G_k + \Sigma\psi_{2,i}Q_{k,i}$$

These last two combinations are significantly less severe than BS 8110 where the imposed loads are significant compared with the dead loads. The use of these lower loadings for some serviceability problems adds to the complexity of the EC2 provisions and to the difficulties in comparing the two codes rigorously. They do, however, have logic on their side. The characteristic loading is generally a loading with a very low probability of actually occurring. It seems reasonable to assume that, for many serviceability problems, exceedance of the criteria very occasionally and for a limited period, is not serious. Problems arise only where the criteria are exceeded frequently or on a fairly continuous basis.

Finally, the two codes are similar in the general level of approach that they deem suitable for most serviceability problems. Both codes rely primarily on the use of 'deemed to satisfy' methods for checking serviceability rather than on explicit calculation of, for example, crack widths. Calculation methods are given but are intended for use only in special cases.

Having made these general points, we can now move on to consider the actual provisions for stress checks, cracking and deflections.

5.2   Stress Checks

EC2 suggests the following limits:

1. $0.6f_{ck}$ under the rare combination of loads where the formation of microcracking or longitudinal cracking could lead to a reduction in durability and where other measures to reduce the problem, such as increasing the cover or providing transverse steel, are not taken.
2. $0.45f_{ck}$ under the quasi-permanent combination of loads if creep is likely to significantly affect the functioning of the member.
3. $0.8f_{yk}$ in ordinary reinforcement or $0.75f_{pk}$ in prestressing tendons under the rare combination of loads. This is to ensure that no inelastic deformations occur in the steel under service conditions.

In general, these limits may be assumed to be met for reinforced concrete provided that the design and detailing for the ultimate limit state are carried out in accordance with Chapters 4.3 and 5 of EC2. It will thus be seen that, in practice, the stress limits may be ignored.

It should be noted that, for prestressed members, the provisions for the control of cracking may impose limits on the tensile stresses in the concrete.

The drafting of this section of EC2 caused a great deal of discussion within the editing group. Many countries, such as Denmark, have never had any requirement to check stresses in their codes and were not prepared to have them introduced now. Other countries, notably those bordering the Mediterranean, still have codes which are basically elastic, permissible stress codes and, while they were happy to have ultimate load methods introduced, they did not wish to lose the elastic stress checks. This explains the rather permissive approach in the drafting which, effectively, permits you to do the checks if you feel they are necessary or to ignore them.

The UK position on this issue is difficult. BS 8110 does not require any checks on stresses for reinforced concrete. Prior to 1957, flexural members were designed on the basis of an elastic analysis of sections and a permissible stress, but columns had been designed on an ultimate load basis since at least 1933, even though the equations had been cast in a pseudo-permissible stress format. In contrast to this, prestressed concrete members are still designed on the basis of permissible stresses under service loads, but with a check for ultimate strength. We thus have no consistent approach to stress checks. The design of prestressed members will be considered in Chapter 7 so will not be considered further here. However, it may be of interest to see what effect the application of stress checks would have on the design of reinforced concrete members, if it were applied.

Figure 5.1 shows, for a rectangular beam section made using concrete with a cube strength of 30 and 460 N/mm^2 reinforcement, the situations where the design will be controlled by the ultimate limit state and where it will be controlled or limited by the compressive stress check. In the derivation of Fig. 5.1 it is assumed that the tension and compression steel provided are fully employed at the ultimate limit state. Since compression steel would not commonly be used where the tension steel was less than about 1.5 per cent, it seems that the stress limitation will only become a controlling factor in normal designs where the steel percentage exceeds about 1.7 and where compression steel is used. Since such sections are relatively uncommon, the compressive stress check would not be likely to be of great practical significance in this case.

The situation for columns is quite different. Figure 5.2 shows a design chart for a column section. The lines on the chart indicate the extra amount of steel needed to meet the compressive stress limit. As shown in section 4.5 the steel areas calculated for ultimate conditions using the two codes will be indistinguishable. It will be seen that the stress check would lead to very much higher areas of reinforcement than is current in the UK for many practical situations. Since there is no evidence of UK designed columns behaving in an unsatisfactory way that could be attributed to excessive compressive stress, it will

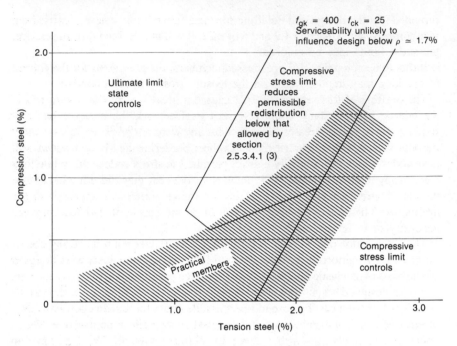

Figure 5.1 Influence of compressive stress checks on design of beams

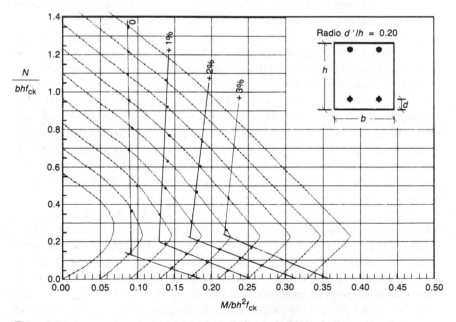

Figure 5.2 Influence of stress checks on column design (only $0.6f_{cu}$ check considered)

be seen that the stress checks should be avoided unless the technical arguments for introducing them were exceptionally strong.

5.3   Control of Cracking

Both EC2 and BS 8110 rely mainly on 'deemed to satisfy' detailing rules for the control of cracking, but give formulae for the prediction of crack widths for use in special cases. In BS 8110, the calculation method is in Part 2, while in EC2 it is in the main body of the text.

As always, EC2 is rather more wordy than BS 8110 but, in this case, this may have resulted in a clearer picture both of the nature of cracking and of the basic principles that lie behind the provisions. EC2 makes the very necessary introductory points that it is effectively impossible to design a structure that can be guaranteed not to crack. It also makes the point that there are many other causes of cracking than just loading or restraint of shrinkage or thermal movements and that control or avoidance of these is outside the control of the designer. It seems very useful to have these points clearly stated in a code.

A guiding principle behind the EC2 rules is that, provided the reinforcement across any possible cracks does not yield, cracking is not likely to be a major problem and control by means of simple detailing rules will be adequate. If the reinforcement does yield then the basic assumption that the reinforcement and the concrete in compression remain elastic will be violated and neither the simple rules nor the calculation method will work. A situation which may lead to yielding of the reinforcement, even where the design for the ultimate limit state has been carried out properly, is where a low percentage of reinforcement has been used and the member is subjected to stresses due to the restraint of imposed deformations. Imposed deformations are deformations imposed on the structure due to effects such as shrinkage or thermal movements. The principle involved can be seen from the following simplified example.

Assume we need to design a 300×300 mm square member to carry a design tension of 100 kN at the ultimate limit state. Using the normal assumptions that the concrete carries no tension leads to a requirement for $100\,000/460 \times 1.15$ mm^2 $= 250$ mm^2 of reinforcement. Assume now that the member was rigidly fixed at its ends and, due to cooling from the temperature generated by hydration, it would shorten by an amount equivalent to a strain of 300×10^{-6}. This is well above the cracking strain of concrete and so a crack can be expected to form. Assuming that the tensile strength of the concrete is 3 N/mm^2, the force generated in the member at the moment of cracking will be $3 \times 300 \times 300N =$ 270 kN. This will be seen to be approaching three times the design yield strength of the reinforcement. The result will be some local yielding of the steel when the first crack forms. No further cracks will be able to form since the force in the member now cannot exceed the force required to yield the steel and the first crack will simply increase in size to accommodate the whole of the temperature movement. The way to avoid this is to include sufficient reinforcement so that it will not yield at first cracking. This will require the provision of sufficient reinforcement to support a tension of 270 kN rather than the design load of 100 kN.

Generalizing this result, it will be seen that, where a member may try to shorten but is restrained from doing so, then controlled cracking can only be achieved if

$$A_c f_t < A_s f_y \quad \text{or} \quad A_s/A_c \geq f_t/f_y$$

More simply, if a member may be subjected to restraint, then at least a minimum area of reinforcement must be provided if controlled cracking is to be ensured. This requirement is stated explicitly in EC2 and a considerable proportion of the clauses dealing with crack control are concerned with ensuring that this minimum is provided. This is done by means of Eq. 4.78 in EC2, repeated below.

$$A_s > k_c k f_{ct.ef} A_{ct}/f_s$$

In this equation A_s is the area of tension reinforcement, A_{ct} the area of concrete within the tension zone, f_s the steel stress which may be taken as the characteristic yield strength of the steel, $f_{ct.ef}$ the tensile strength of the concrete effective at the time the first cracks form and k_c, k the coefficients depending on the nature of the member and of the restraint. This will be seen to be a more generalized form of the equation derived above.

While BS 8110 does not use such a highly developed approach for defining minimum areas of reinforcement nor define its objectives so clearly, it does include rules which aim to achieve the same objectives. These rules are given in clause 3.12.5.3 and are therefore not so clearly associated with crack control. The *Handbook* to BS 8110 does, however, make clear that this is one of their purposes. Table 5.1 gives a comparison of the minimum percentages required by BS 8110 and typical values from EC2, taking $f_{ct.ef}$ as 3 N/mm² (the minimum value suggested in EC2). It will be seen that the codes do not differ by too much in the particular cases considered.

The 'deemed to satisfy' detailing rules in BS 8110 and EC2 are both developed from parameter studies carried out using the respective crack prediction formulae. It therefore seems helpful to compare these formulae before attempting to compare the simple rules. Table 5.2, taken from the studies carried out for BRE by BCA, presents a direct comparison of the two sets of provisions.

Table 5.1 Minimum reinforcement provisions for crack control

Member type	Code	Minimum percentage
Element subject to pure tension, thickness 300 mm	BS 8110	0.45
	BS 8007	0.35
	EC2	0.52
Element subject to pure tension, thickness 800 mm	BS 8110	0.45
	BS 8007	0.22
	EC2	0.33
Rectangular beam	BS 8110	0.13
	EC2	0.10

At first glance there is little similarity between the two sets of provisions. In practice, however, they are not so different and probably the most significant differences are those noted at the bottom of Table 5.2 The difference in the probabilities of exceedance will, on average, lead to the EC2 formula predicting a width about 10 per cent greater than BS 8110. Against this, EC2 requires the check to be done under significantly lower loads. To gain some idea of the reduction implied, the quasi-permanent load is given by

$$\text{Quasi-permanent load} = G_k + \psi_2 Q_k$$

Reasonable values for ψ_2 would seem to be domestic occupation 0.2, offices and stores 0.3, parking 0.6.

The assumption of a value of 0.3 for ψ_2 gives the following ratios of design loads for various ratios of characteristic imposed to permanent loads:

Q_k/G_k	EC2 load/BS 8110 load
1.2	0.62
1.0	0.65
0.5	0.77
0.2	0.88

For normal structures, it would seem that the EC2 loads are likely to be about 30 per cent below those used in BS 8110.

Table 5.2 Comparison of BS 8110 and EC2 crack formulae

BS 8110	EC2
$w = \dfrac{3a_{cr}\epsilon_m}{1+2(a_{cr}-c)/(h-x)}$ This reduces to $w = 3a_{cr}\epsilon_m$ for pure tension	$w = 1.7(50+k\phi/\rho_r)\epsilon_{sm}$ $k = 0.1$ for flexure $ 0.2$ for tension
(a) For flexure: $\epsilon_m = (\epsilon_1 - b(h-x)^2/3(d-x)A_s E_s$ (b) For tension: $\epsilon_m = \epsilon_1 - 2bh/3E_s A_s$	$\epsilon_{sm} = \epsilon_1(1 - 0.5(\sigma_s/\sigma_{sr})^2)$
Formula predicts width with a 20% chance of exceedance Calculation is done for member carrying full service load Strain used is that at the tension face	Formula predicts width with a 5% chance of exceedance Calculation is done for carrying quasi-permanent load Strain is that at the steel level

The notation is as follows: a_{cr} is the distance from the point considered to the surface of the nearest bar, c the cover, h the overall depth of section, x the neutral axis depth, w the crack width, k a coefficient, b the section breadth, E_s the elastic modulus of reinforcement, A_s the area of tension steel, ϵ_1 the strain in reinforcement calculated on basis of a cracked section, ϵ_{sm} the average strain in reinforcement, ϵ_m the average strain at level considered, σ_{sr} the stress in steel calculated on basis of a cracked section at the cracking moment, σ_s the stress in steel under moment considered calculated on the basis of a cracked section, ρ_r the effective reinforcement ratio and ϕ the bar diameter.

Using the strain at the level of the centroid of the tension steel rather than at the tension face is likely to lead to EC2 using strains that are 10–25 per cent below those used by BS 8110.

Taking these three factors together, it will be seen that EC2 could be expected to give design crack widths about 30–40 per cent smaller than BS 8110, even if the formulae were otherwise identical.

It is not possible to make a rigorous, simple comparison between the two sets of formulae, but Table 5.3, which presents results from parameter studies carried out on beam sections for BRE, should give a general picture. In the study, it was assumed that the quasi-permanent load was 75 per cent of the service load as used by BS 8110.

Table 5.3 Comparison of calculated crack widths for beams

Breadth (mm)	Overall depth (mm)	Tension steel (mm)	Cover (mm)	Crack width (mm)		
				EC2	BS 8110	EC2/BS 8110
200	300	2–25	30	0.12	0.20	0.60
400	300	2–25	30	0.18	0.30	0.60
600	300	2–25	30	0.24	0.35	0.69
285	200	2–25	30	0.10	0.22	0.45
285	400	2–25	30	0.18	0.29	0.62
285	600	2–25	30	0.20	0.32	0.63
285	300	2–16	30	0.21	0.25	0.84
285	300	2–20	30	0.18	0.27	0.67
285	300	2–25	20	0.14	0.24	0.58
285	300	2–25	40	0.14	0.30	0.47
285	300	2–25	50	0.14	0.32	0.44

The relative performance of the formulae for slabs is rather different, as can be seen from Fig. 5.3 which compares the calculated crack widths in a 300 mm deep slab reinforced with 20 mm bars at different spacings. Here the results are rather more similar. The same is true for members loaded in tension.

Pure tension resulting from loading is not common. A much more common situation is where cracking results from tension induced by the restraint of imposed deformations. These imposed deformations may arise from shrinkage or from thermal movements. For this situation EC2 states that σ_s may be taken as equal to σ_{sr}. This simplifies the equation for the average strain to

$$\epsilon_{sm} = 0.5\sigma_{sr}/E_s$$

σ_{sr} will be proportional to the tensile strength of the concrete.

This leads to the apparently odd result that the crack width is constant and independent of the magnitude of the imposed deformation. In fact, the result is reasonable and can be derived from the following considerations. Consider a member as shown in Fig. 5.4 which is rigidly restrained at its ends and is cooled down. As the member cools it tries to shorten but, since it is restrained, it is subjected to a tensile stress. Eventually, this stress will reach the tensile strength

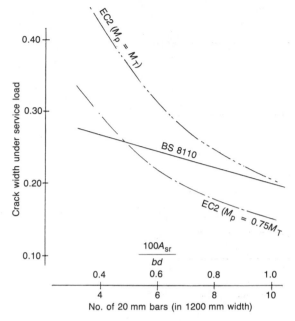

Figure 5.3 Comparison of calculated crack widths in slabs. Constant bar size, varying spacing

of the concrete and the first crack will form. The formation and opening of this crack absorbs some of the shortening and the stress is reduced. Further cooling causes the stress to increase again until once more it reaches the tensile strength of the concrete and a second crack forms, leading to further stress relief. This process is repeated until cooling stops or a maximum possible number of cracks has formed. During the whole of this process, the force in the member never exceeds that corresponding to the tensile strength of the concrete. This constant maximum force corresponds to a constant crack width. This process is illustrated schematically in Fig. 5.5. This approach will be seen to be very different to that assumed in British codes where the crack width is assumed to be directly proportional to the restrained strain. The predictions of EC2 and British codes will thus be very different for this type of problem.

There is one further problem with the application of EC2 which has not been mentioned. In order to calculate σ_{sr} or the minimum steel percentage, it is necessary to have a value for the tensile strength of the concrete. EC2 gives three possible values for this: a mean value, an upper characteristic and a lower characteristic strength. It is not specific about which is the more appropriate value to choose except that it is stated that, for calculating the minimum steel percentage, one should not take a value < 3. For crack widths due to loading, the BRE study concluded that closest agreement with BS 8110 was obtained by taking the lower characteristic tensile strength. For cracks caused by restrained imposed deformations, logic would suggest use of the upper characteristic strength since this would give the worst case. No useful comparison with UK practice can really

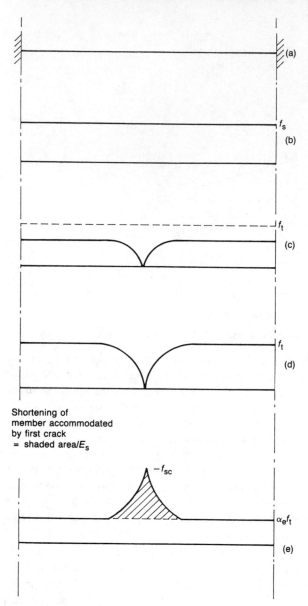

Figure 5.4 Stress conditions in the early phases of cracking due to restrained shortening. (a) Restrained member; (b) stress in concrete just before first crack; (c) stress in concrete just before second crack; (d) stress in steel just before second crack

be made in this case to determine which strength gives the best agreement. Possibly the best solution would be to take the value of 3 N/mm² specified for calculation of the minimum steel areas.

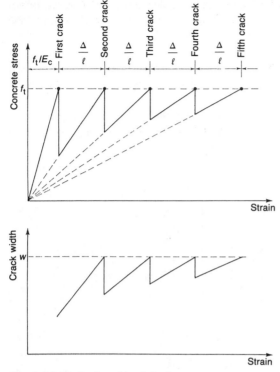

Figure 5.5 Idealized cracking behaviour

5.4   Simplified Rules for Crack Control

Both BS 8110 and EC2 give simplified detailing rules for ensuring crack control in reinforced concrete so that generally it is unnecessary to calculate crack widths explicitly. The methods in the two codes are formulated differently and so cannot be compared directly. However, it is possible to reformulate the BS 8110 equation so that it is in a similar form to the EC2 limits and this will allow a more direct comparison to be made.

The two sets of provisions are as follows for reinforced concrete:

1   Depth of member below which no action is required to control cracking.
    BS 8110: no check if $h < 250$ mm and $f_y = 250$ or if $h < 200$ mm and
    $f_y = 460$ or if $100A_s/bd < 0.3$
    EC2: no check if $h < 200$ mm
2. Steel stress considered.
    BS 8110: steel stress calculated on the basis of a cracked section under the characteristic load. This may be assumed to be

$$\frac{5}{8}f_y A_{s.\text{req}}/A_{s.\text{prov}}$$

EC2: steel stress calculated on the basis of a cracked section under the quasi-permanent load. No simplified rule is given for the estimation of the steel stress.

3. Detailing requirements.

BS 8110: in beams the clear spacing between bars $\leq 47\,000/f_s$; in slabs if $0.3 < 100A_s/bd < 1$ the spacing given for beams may be divided by $100A_s/bd$.

EC2: For the estimated steel stress, either the maximum bar size or the maximum bar spacing, whichever is the more convenient, may be read off Fig. 5.6.

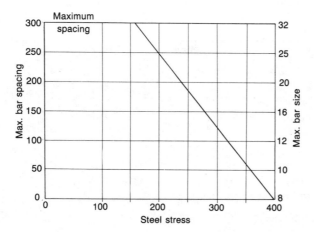

Figure 5.6 Maximum bar spacing or diameter rules in EC2

The important point to note about the EC2 detailing rules is that the user may either choose to limit the bar spacing or to limit the bar diameter. Effectively, he may choose the least critical option. This is often misunderstood by readers of EC2.

The BS 8110 formula for bar spacing can be rewritten in the following way. For a slab with a single layer of tension steel, the reinforcement percentage can be expressed as

$$\rho = 100A_s/bd = 100 \times \pi \times \phi^2/(4ds)$$

where s is the centre-to-centre spacing of the bars. Hence the expression for bar spacing can be rewritten as

$$s_c = 4 \times 470sd/(\phi^2 f_s)$$

If it is assumed that s_c is approximately equal to s, then the spacing can be eliminated from the equation, which can then be rearranged to give

$$\phi = 24.5(d/f_s)$$

Thus the BS 8110 rule for slabs can be reformulated as a limitation on bar diameter rather that on bar spacings. (This derivation ignores the bottom limit to the steel percentage of 0.3.) Assuming that the steel stress can be taken to be $\frac{5}{8}f_y$ and assuming a section depth of 300 mm, then for $f_y = 460$, the above relationship gives a maximum bar diameter of 25 mm. For steel percentages > 1, the normal formula gives a clear spacing of 165 mm. Using the relationship derived above for the relation between bar size, steel percentage and bar spacing, it will be found that, for percentages > 1, the bar diameter will exceed 25 mm if the bar spacing is 165 mm. For steel percentages < 1, the bar spacing will exceed 165 mm if a bar size of 25 mm is used. Thus, for the example given above, crack control could have been achieved by ensuring that, for any steel percentage, either

$$\phi \leq 25 \quad \text{or} \quad s \leq 165$$

This discussion shows that, despite the very different formulation of the 'deemed to satisfy' rules in the two codes, they are, in principle, the same. On the face of it, the only significant difference is that the BS 8110 bar diameter rule is a function of section depth. Even here, however, there is a similarity since, in a note below Table 4.10 in EC2, it is stated that the maximum bar size may be increased by a factor $0.1h/(h-d)$. Here $(h-d)$ is the cover plus half the bar size and so, for a given cover, the maximum bar size is proportional to the section depth. This compares with the reformulated BS 8110 rule where the diameter is proportional to the square root of the effective depth.

Having reformulated the BS 8110 provisions, it is now possible to compare them directly. This is done in Table 5.4 for a 250 mm deep slab using grade 460 steel. It must be remembered that, since EC2 requires cracks to be checked under the quasi-permanent loads while BS 8110 checks under the full service loads, the steel stress used in the two codes for identical slabs will be different.

Comparison of the tables shows that at low stress levels, EC2 permits both wider bar spacings and larger bar sizes than does BS 8110. This is consistent with the studies presented earlier in the chapter which showed that EC2 was generally less conservative than BS 8110. At higher stress levels, however, the opposite is true and BS 8110 is the more permissive. It should be noted, however, that these stress levels are very high. Under the full service load, BS 8110 only

Table 5.4 Comparison of simplified rules for crack control

Steel stresses (N/mm²)		Maximum spacing (mm)		Maximum bar size (mm)	
EC2	BS 8110 equivalent	EC2	BS 8110	EC2	BS 8110
160	229	300	230	32	24
200	286	250	190	25	21
240	343	200	160	20	19
280	400	150	135	16	18
320	457	100	115	12	17
360	514	50	95	10	16

gives wider spacings or diameters at stresses of 400 N/mm^2 or above. At first sight these stress levels look unrealistically high; however, it needs to be remembered that EC2 is concerned with the control of cracks due to restrained imposed deformations as well as those due to loading and that EC2 permits stresses up to the yield of the steel in this case. The BS 8110 rules are not intended to apply in this situation, though it should be noted that, effectively, BS 8110 does allow stresses up to yield to be used in defining the minimum steel percentages.

In summary, the crack control provisions of the two codes are not as different as might at first appear; however, due mainly to the different levels of loading under which the checks are carried out, EC2 will generally be somewhat less restrictive than BS 8110. It should be remembered that EC2 is much more concerned about cracking due to the restraint of imposed deformations than about load-induced cracking. BS 8110 does not directly deal with this problem in Part 1, while the approach used in Part 2 or in BS 8007 is quite different. The EC2 approach seems to be the more logical.

5.5   Control of Deflections

Like crack control, deflections are dealt with in both codes by simple 'deemed to satisfy' rules in the main body of the code, but a direct calculation method is also given. In BS 8110 the calculation method is given in Part 2 while in EC2 it is in Appendix 4. In both codes, the 'deemed to satisfy' approach is based on span/effective depth ratios. These have, in both cases, been derived from parameter studies using the calculation methods. Thus, even though design will not often use the calculation methods directly, it seems most appropriate to compare these first.

5.5.1   BS 8110 Method for Deflection Calculation

BS 8110 provides a series of assumptions which are sufficient to define the behaviour of a section under any condition of loading. These assumptions are as follows:

1   If the section is uncracked then both the steel and the concrete are assumed to behave elastically both in tension and compression.

2.   If the section is cracked then the reinforcement is assumed to behave elastically in both tension and compression. The concrete is assumed to behave elastically in compression but, in tension, a triangular distribution of stress is assumed over the tension zone with zero stress at the neutral axis and a tensile stress of f_t at the level of the centroid of the tension steel. In the short term, f_t takes a value of 1 N/mm^2 and, in the long term, 0.55 N/mm^2. Plane sections are assumed to remain plane.

Whether the section is cracked or uncracked is simply decided by taking whichever set of assumptions gives the greater deflection. The assumptions are

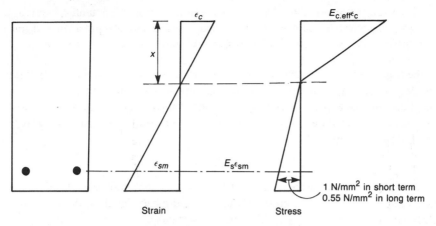

Figure 5.7 BS 8110 assumptions for calculating curvatures or a cracked section

illustrated in Fig. 5.7. It will be seen that the method effectively assumes a tensile strength of concrete which is slightly greater than 1 N/mm² in the short term and reduces to a value slightly greater than 0.55 N/mm² in the long term. The significance of this low value will be discussed further when the comparisons with EC2 are discussed. Calculation of the curvature of a section can only be done iteratively, which is no great problem for a computer, but is very tedious by hand.

Having calculated the curvature, the deflection may be calculated either by double numerical integration of the curvatures along the beam or by use of simplified methods based on deflection coefficients. This second method will be conservative.

Shrinkage curvature is calculated from the relation

$$(1/r)_{sh} = \epsilon_{sh} \alpha_e S/I$$

where $(1/r)_{sh}$ is the curvature due to shrinkage, ϵ_{sh} the free shrinkage strain, α_e the effective modular ratio, allowing for creep, S the first moment of area of the reinforcement about the centroid of the section and I is the second moment of area of the section. Here S and I are calculated either on the basis of an uncracked or a fully cracked section, ignoring concrete in tension, whichever is the more appropriate.

5.5.2   EC2 Method for Deflection Calculation

EC2 takes as its starting-points for the calculation of the curvature of a section under load two limiting conditions:

1   The uncracked state. The curvature in the uncracked state is calculated exactly as in BS 8110.

2. The fully cracked state. The curvature in this state is calculated assuming that the reinforcement is elastic in both tension and compression and that the concrete is elastic in compression. Concrete in tension is ignored.

If the tensile stress at the tension face of the section is below the tensile strength of the concrete, f_{ctm}, then the section is assumed to be uncracked. Above this level the curvature lies between that calculated on the basis of the fully cracked and that calculated on the basis of the uncracked section. This is calculated from the following relationship.

$$(1/r) = \xi(1/r)_1 + (1-\xi) \times (1/r)_0$$

where $(1/r)$ is the curvature of the section, $(1/r)_0$ the curvature calculated on the basis of an uncracked section, $(1/r)_1$ the curvature calculated on the basis of a fully cracked section and ξ a distribution coefficient.

The coefficient ξ is a function of how close conditions are to the condition causing cracking. It takes a value of zero at the cracking moment and approaches unity as the loading increases above the cracking moment. It is given by the relation

$$\xi = 1 - \beta_1\beta_2(\sigma_{sr}/\sigma_s)^2$$

where β_1 is a coefficient which takes account of the bond properties of the reinforcement, it takes a value of 1 for high bond bars and 0.5 for smooth bars; β_2 is a coefficient which takes account of the duration and nature of the loading, it takes a value of 1.0 for short-term loads and 0.5 for sustained or cyclic loads; σ_{sr} is the stress in the tension steel calculated on the basis of a fully cracked section under the loading which will just cause cracking at the section being considered; σ_s is the stress in the tension steel calculated on the basis of a cracked section under the loading for which the curvature is being calculated. The curvature due to shrinkage is calculated using the same formula as in BS 8110, but for a cracked section in EC2 the curvatures in both the uncracked and the fully cracked state are calculated and the appropriate value assessed using the distribution coefficient.

5.5.3   Comparisons of the EC2 and BS 8110 Methods

Figure 5.8 shows a parametrized moment—curvature diagram for a rectangular section having 0.5 per cent of tension reinforcement calculated using EC2 and BS 8110. It will be seen that the curves are generally similar but that, at values of M/bd^2 between 0.3 and 0.6, EC2 gives substantially lower curvatures than BS 8110. This arises because, as mentioned earlier, BS 8110 effectively uses a low tensile strength for the concrete of around 1. EC2 uses a much higher value. While it can be argued that the EC2 value is much more realistic for the tensile strength, it can also be argued that the tensile strength is often effectively reduced by restrained shrinkage or other factors and that the BS 8110 approach is therefore the more prudent one.

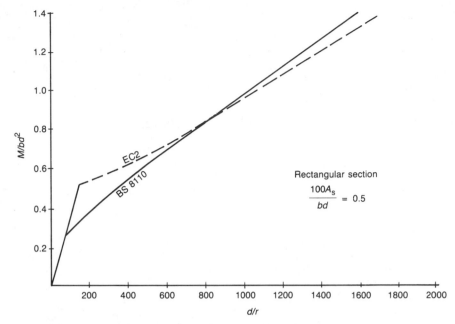

Figure 5.8 Comparison of BS 8110 and EC2 methods of curvature calculation

A more general comparison of the methods is given in Fig. 5.9. This shows the parametrized curvature for rectangular sections subjected to two-thirds of their ultimate loads (roughly the service load) plotted against the reinforcement percentage. It will be seen that the two curves really only differ significantly below about 0.25 per cent. This suggests that, in practical terms, the two methods are more or less equivalent.

While the effect of the two methods on the resulting design may be roughly equivalent, the EC2 method does have one major practical advantage over BS 8110 for most common problems. This is that it is much easier to apply since it does not require iterative calculations, it simply requires that the properties of the uncracked and the fully cracked section and the cracking moment can be found. All these can be obtained from standard formulae or from design charts.

The EC2 method does, however, have one disadvantage which could cause problems in the treatment of members subjected to axial loads as well as bending. This is the problem of defining the cracking moment since this will depend on how the loads are applied. If the axial force is applied first and then the bending, the cracking moment will be very different from what would be obtained if the bending was applied first and then the axial load or if the bending moment and axial load were increased together. EC2 gives no guidance on how to handle this problem, though, for prestressed members, it is clear that the axial load is applied before the bending. This problem does not arise with the BS 8110 method since the cracking moment is not an essential parameter in the calculation.

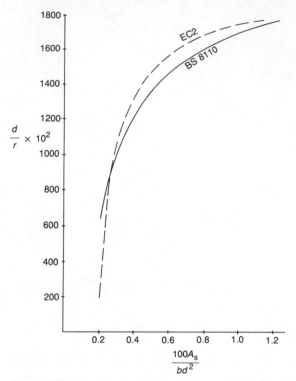

Figure 5.9 Relation between curvature under service load and steel percentage

5.5.4   Simplified Methods

The simplified methods used for checking deflections in the two codes are essentially the same as they both employ limits to the ratio of the span to the effective depth. The method given in EC2 is rather more simplified than that in BS 8110. This makes EC2 less flexible and, in some practical situations, considerably more conservative than BS 8110.

The EC2 method is set out in Table 5.5. This applies to beams or slabs designed using reinforcement with a characteristic strength of 400 N/mm^2. For other characteristic strengths, the values in Table 5.5 may be multiplied by $400/f_y$. 'Concrete highly stressed' corresponds to sections having 1.5 per cent or more of tension reinforcement and 'concrete lightly stressed' corresponds to sections with 0.5 per cent or less of tension steel. Interpolation between these cases is permitted. For flanged sections the values in Table 5.5 are multiplied by 0.8.

Table 5.6 compares the span/effective depth ratios specified by the two codes for the cases considered in EC2 where 460 N/mm^2 reinforcement is used.

It will be seen that agreement at these points is very close with EC2 being marginally more conservative, except for cantilevers. The most significant difference in Table 5.6 is for waffle slabs. In the interpretation of BS 8110, it

Table 5.5 EC2 Span/effective depth ratios

Structural system	Concrete highly stressed	Concrete lightly stressed
1. Simply supported beam, one of two-way spanning simply supported slab	18	25
2. End span of continuous beam or one-way continuous slab or two-way slab continuous over one long side	23	32
3. Interior span of beam or slab	25	38
4. Flat slab (based on longer span)	21	30
5. Cantilever	7	10

Table 5.6 Comparison of BS 8110 and EC2 span/depth ratios

Structural system	$\rho = 0.5\%$ $M/bd^2 = 1.8$		$\rho = 1.5\%$ $M/bd^2 = 4.75$		Section shape
	BS 8110	EC2	BS 8110	EC2	
1.	22.5	21.5	16.5	15.5	Rect.
	18.0	17.2	13.2	12.4	T
2.	29.3	27.5	21.5	19.8	Rect.
	23.4	23.9	17.2	17.4	T
3.	29.9	29.9	21.5	19.8	Rect.
	23.4	23.9	17.2	17.4	T
4.	26.3	25.1	19.3	18.6	Solid
	23.4	20.1	17.2	14.9	Waffle
5.	7.9	8.4	5.8	6.2	Rect.
	6.3	6.7	4.6	5.0	T

has been assumed that there is a solid section of slab in the region of the column sufficiently large for it to be considered as a drop. This permits the factor of 0.9, given in BS 8110 for slabs without drops to be neglected. EC2 does not introduce any such refinements.

Table 5.6 does not give a complete picture of the comparison of the two approaches as it only makes comparisons of cases with 0.5 and 1.5 per cent of reinforcement. Figure 5.10, drawn for simply supported beams, gives a more complete picture of how the codes compare over the full practical range of reinforcement ratios. EC2 defines just two points and permits interpolation between these. BS 8110 provides a continuous curve. It will be seen from the Fig. 5.10 that agreement between the codes is very close for values of $M/bd^2 > 1.8$, corresponding to a steel percentage of 0.5. For lower percentages, however, EC2 becomes increasingly conservative relative to BS 8110 due to the cut-off imposed. It is interesting that this conservatism occurs in just that region where the EC2 calculation method is *less* conservative than BS 8110. This region is also one of great practical significance since it will include many, if not most, slabs. There would thus seem to be very significant gains to be made if the deflections in slabs were checked by calculation when using EC2, both over the EC2 span/depth ratio method and over BS 8110.

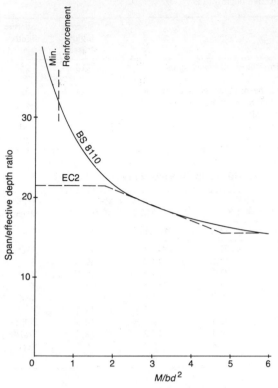

Figure 5.10 Comparison of span/effective depth ratios for simply supported rectangular sections $(f_y = 460)$

Both EC2 and BS 8110 require the use of a reduced span/effective depth ratio for longer spans. In EC2 the ratio is multiplied by the factor, 8.5/span for flat slabs and 7/span for other members when the span exeeds 8.5 or 7 m respectively. In BS 8110, the ratio is multiplied by 10/span for spans greater than 10 m where there are finishes or partitions which are liable to be damaged by deflection. In this respect, EC2 will be seen to be more conservative than BS 8110. In BS 8110 this factor is introduced to satisfy a maximum deflection limit of 20 mm where partitions or finishes might be damaged. Interestingly, EC2 introduces no such constant maximum deflection limit.

6 Detailing

6.1 General

Although the objective of the detailing rules in EC2 are similar to that of BS 8110, they are expressed differently and in most situations provide different rules. The approach is generally logical and for most normal situations the quantity of reinforcement will be similar. The layout of the detailing is different. EC2 starts with spacing rules and minimum radius of bends followed by general clauses on bond and achorage. The detailing rules are then given for each type of element.

Much more attention is given to detailing with welded meshes. This reflects the much greater use of these in other countries (especially Germany) both for slabs where flat sheets are used and for beams where the meshes are bent to form cages.

BS 8110 has a section covering the design and detailing of ties. This is reflected in EC2 only as general statements. It does not include detailed clauses.

Curtailment of reinforcement for beams and slabs in EC2 requires the bending moment envelope to be shifted to account for the chord tension implied by the truss analogy used for shear design. This increases the length of bar by about $d/2$ for beams and d for slabs. This is in addition to the normal bond length which is similar to that of BS 8110.

Straight anchorage of plain bars with diameter of more than 8 mm is not permitted by EC2. All plain bars must be bent or hooked to obtain full anchorage.

Bends and hooks are not recommended for compression anchorage. The main clauses of EC2 are discussed below and compared with BS 8110. EC2 clause numbers are shown in parentheses with subheadings.

6.2 Spacing of Bars (clause 5.2.1.1)

EC2 specifies the clear distance between bars, both horizontal and vertical. Between individual parallel bars or horizontal layers of parallel bars it should not be less than *maximum bar diameter or 20 mm*. In addition where the maximum aggregate size, d_g, is > 32 mm these distances should not be less than d_g or the bar size, whichever is greater.

BS 8110 requires that the horizontal distance between bars should not be less than $h_{agg} + 5\,mm$, where h_{agg} is the maximum size of the coarse aggregate. Where there are two or more rows the following should apply:

1. The gaps between corresponding bars in each row should be vertically in line;
2   The vertical distance between bars should not be less than $2h_{agg}/3$. When the bar size exceeds $h_{agg} + 5\,mm$, a spacing less than the bar size or equivalent bar size should be avoided.

This difference between the two codes could show itself where the aggregate size was 20 mm and the bar size was 16 mm. EC2 would require the clear vertical distance to be 20 mm and BS 8110 would require it to be 16 mm.

6.3   Permissible Radius of Bends (clause 5.2.1.2)

EC2 specifies the minimum diameter of mandrel round which reinforcement should be bent. It is easy to confuse this with the minimum radius of bend referred to in BS 8110. BS 8110 assumes that the minimum radius of bends are as given in BS 4466. The values given in the NAD for EC2 correspond to those.

EC2 limits the radius of bend with respect to the side cover by a simple tabular means. This corresponds to Eq. 50 in BS 8110. For the values given in Table 6.1 the codes give similar results ($f_y = 460$, $f_{cu} = 30$), as shown. EC2 provides special rules for bending meshes, see Table 6.2. There is no equivalent in BS 8110. These rules should be used when forming beam cages from welded mesh.

Table 6.1   Minimum diameter of mandrel for a given side cover

Side cover (mm)	EC2	BS 8110
>100 mm and >7ϕ	10ϕ	13ϕ
>50 mm and >3ϕ	15ϕ	15.7ϕ
≤50 mm and ≤3ϕ	20ϕ	20.9ϕ

Table 6.2   Minimum diameter of mandrel for welded mesh to EC2

	Minimum diameter of mandrel	
Welds outside bends		Welds inside bends

$d < 4\phi$: 20ϕ

$d \geq 4\phi$: Table 6.1 applies

20ϕ

6.4 Bond (clause 5.2.2)
6.4.1 Bond Condition

EC2 describes bond as 'good' or 'poor' (clause 5.2.2.1). The angle of the bar to the direction of pouring concrete is considered an important factor. Horizontal bars which are placed in the top of a section which is more than 250 mm deep are considered to have poor bond (see Fig. 6.1). BS 8110 has a less conservative rule (clause 3.12.8.13) which applies only to the design of tension laps. This rule in EC2 provides an unfortunate step function. The top steel in a slab of 275 mm depth has 'poor' bond, whereas that of 250 mm depth has 'good' bond. 'Poor' bond conditions incur a penalty factor of 0.7 on the permissible bond stress.

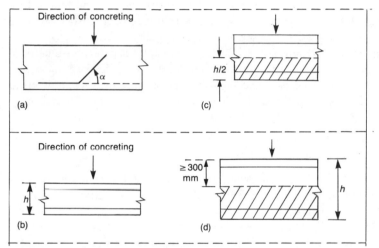

Figure 6.1 Definitions of bond conditions. (a) $45° \leq \alpha \leq 90°$ for all h values; (b) $h = 250$ mm; (c) $h = 250$ mm; (d) $h = 600$ mm. (a) and (b) Good bond conditions for all bars. (c) and (d) Bars in hatched zone: good bond conditions; bars not in hatched zone: poor bond conditions

Values of design bond stress in EC2 are very similar to those of BS 8110 for normal grades of concrete, but as the strength of concrete increases EC2 provides an increase of bond stress as shown in Table 6.3. Enhancement of bond stress is permitted in EC2 where transverse pressure exists (clause 5.2.2.2(3)). This is limited to $1/(1 - 0.04p) \not> 1.4$, where p is the mean transverse pressure (N/mm^2). No such increase in bond stress is given in BS 8110.

Table 6.3 Comparison of design bond stress for high bond bars

f_{ck}/f_{cu}	20/25	25/30	30/37	35/45	40/50	45/55	50/60
EC2	2.3	2.7	3.0	3.4	3.7	4.0	4.3
BS 8110	2.5	2.74	3.0	3.4	3.5	3.7	3.9

6.4.2   Basic Anchorage Length (clause 5.2.2.3)

EC2 and BS 8110 both assume a constant bond stress along the length of a bar. The basic anchorage length is thus

$$l_b = (\phi/4)(f_{yd}/f_{bd}) \tag{6.1}$$

where ϕ is the diameter of bar, f_{yd} the design strength of bar and f_{bd} the design bond stress.

6.5   Anchorage (clause 5.2.3)

EC2 states in the principles that the anchorage of bars must avoid longitudinal cracking or spalling of the concrete. Transverse reinforcement must be provided where necessary. BS 8110 does not state this explicitly.

Special attention is required in EC2, where mechanical devices are used, to check their capacity to transmit the concentrated force. EC2 does not permit straight anchorages or bends to anchor plain bars of more than 8 mm diameter (clause 5.2.3.2). It does not recommend the use of bends or hooks for anchorage of bars in compression. BS 8110 does not contain such special clauses.

6.5.1   Transverse Reinforcement

EC2 recognizes the pinching effect on the anchorage of the main tension bars at the support of a beam (clause 5.2.3.3). Where this does not exist (e.g. indirect supports) transverse reinforcement should be provided. The area of this transverse reinforcement should be at least one-quarter of the area of the anchored steel and evenly spaced along its length. BS 8110 always requires links but does not state this reason for their need.

In EC2, Fig. 5.3 implies that transverse reinforcement in slabs should be placed outside the anchored bars. However, this is not stated. BS 8110 does not require this and it is not common practice to do so. It is more usual to place the main steel in slabs in the outer layer.

6.5.2   Required Anchoraged Length (clause 5.2.3.4)

EC2 introduces a number of factors which modify the basic anchorage length, l_b, and sets a minimum length, $l_{b,min}$, of

$$0.3l_b \ (\not< 10\phi) \text{ in tension} \quad \text{and} \quad 0.6l_b \ (\not< 100\phi) \text{ in compression}$$

These minimum values differ from BS 8110 which requires the greater of d or 12ϕ (clause 3.12.9.1). In addition BS 8110 requires that for bars anchored in the tension zone, one of the following distances for all arrangements of design ultimate load should be considered:

1. An anchorage length appropriate to its design strength ($0.87f_y$) from the

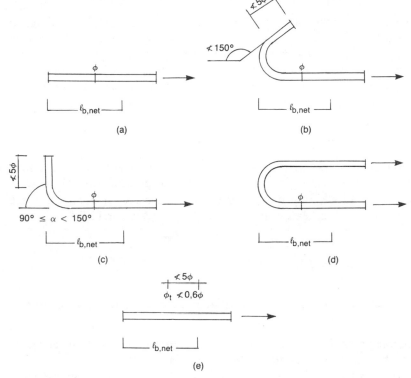

Figure 6.2 Required anchorage length. (a) Straight anchor; (b) hook; (c) bend; (d) loop; (e) welded transverse bar

point at which it is no longer required to assist in resisting the bending moment; or

2. To the point where the design shear capacity of the section is greater than twice the design shear force at that section; or

3. To the point where other bars continuing past that point provide double the area required to resist the design bending moment at that section.

In EC2 the required anchorage length, $l_{b,net}$ is calculated from

$$l_{b,net} = \alpha_a l_b \frac{A_{s,req}}{A_{s,prov}} \geq l_{b,min} \qquad [6.2]$$

where l_b and $l_{b,min}$ are defined as above, $A_{s,req}$ and $A_{s,prov}$ the areas of reinforcement required and provided respectively and α_a a coefficient which takes the following values: 1 for straight bars, 0.7 for curved bars in tension (see Fig. 6.2) if concrete cover perpendicular to the plane of curvature is at least 3ϕ in the region of the hook, bend or loop.

6.5.3   Hooks and Bends

The consequence of applying Eq. 6.2 is that the anchorage allowance for standard hooks and bends is different in the two codes. EC2 reduces the required anchorage length by 0.7 regardless of what sort of bend (90° or hook). BS 8110 assumes an equivalent length of straight bar for hooks and bends. EC2 requires that the bar extends at least 5ϕ beyond the bend. BS 8110 requires that the bar extends at least 4ϕ beyond the bend.

6.5.4   Anchorage of Welded Meshes

EC2 reduces the anchorage length by 0.7 for welded meshes (clause 5.2.3.4.2). Where a welded transverse bar exists in the anchorage zone, BS 8110 reduces the anchorage length only if the number of welded intersections within the anchorage length is at least equal to $4A_{s,prov}/A_{s,rqd}$ (clause 3.12.8.5).

6.6   Splices (clause 5.2.4)
6.6.1   Lapped Bars

EC2 has very specific rules concerning the position and spacing of lapped bars (clause 5.2.4.1.1, Fig. 5.4).

1.   Lapping bars should not be further apart than 4ϕ (otherwise the lap length should be increased by the amount exceeding 4ϕ).
2.   Laps should be staggered with at least $0.3 \times$ lap length measured longitudinally between adjacent laps.
3.   The pitch lapped bars should be at least 2ϕ or 20 mm whichever is greater.

BS 8110 does not have any equivalent clauses although rules are given to control minimum clear distance between adjacent laps (clause 3.12.8.13).

6.6.2   Transverse Reinforcement (clause 5.2.4.1.2)

EC2 is more specific than BS 8110 concerning transverse reinforcement at splices (clause 5.2.4.1.2).

For bar sizes 16 mm and above the total area of transverse reinforcement in the length of the splice should be not less than the area of one spliced bar. Figure 5.5 of EC2 suggests that the transverse reinforcement should be bunched at each end of the splice. This transverse reinforcement should be in the form of links if the clear distance between lapped bars is $\leq 10\phi$ (NAD) and be straight in other cases. The transverse reinforcement should be placed between the surface of the concrete and the longitudinal steel. The NAD makes it clear that this only applies to beams.

In contrast, BS 8110 only provides rules for minimum area of links in beams and columns (clause 3.12.7), minimum area of transverse reinforcement in flanges of flange beams (Table 3.27) and a minimum spacing of bars in slabs (clause 3.12.11.7). There is no requirement that transverse reinforcement in slabs should be between the surface of the concrete and the longitudinal steel.

6.6.3   Lap Length (clause 5.2.4.1.3)

EC2 bases the lap length, l_s, on the required anchorage length. It must not be <0.3 times the basic anchorage length, l_b, or 14ϕ or 200 mm.

Where more than 30 per cent of the bars in the section are lapped and the clear distance between lapped bars is $< 6\phi$ (NAD) or the side cover is $< 2\phi$ (NAD), tension lap lengths are increased by a factor of 1.4. If both these conditions prevail tension laps are increased by a factor of 2.

BS 8110 has similar rules to control the length and amount of lapped bars (clauses 3.12.8.11−14). The minimum lap length for bar reinforcement should not be less than 15 times the bar size or 300 mm, whichever is greater. At laps, the sum of reinforcement sizes in a particular layer should not exceed 40 per cent of the breadth of the section at the level. Where a tension lap occurs at the corner of a section and the minimum cover to either face is less than twice the size of the lapped reinforcement, or where the clear distance between adjacent laps is less than 75 mm or six times the size of the lapped reinforcement whichever is the greater, the lap length should be increased by a factor of 1.4.

BS 8110 provides further limits for lapped bars (in tension) cast in the top of beam and slabs where the minimum cover is less than twice the size of the lapped reinforcement, the lap length should be increased by a factor of 1.4. Where both this condition exists and the spacing of bars is less than the limit given above then the lap length should be increased by a factor of 2.

6.6.4   Laps for Welded Mesh Fabrics of High Bond Wires (clause 5.2.4.2)

EC2 contains more detailed clauses than BS 8110. This is not surprising since more use of meshes is made outside the UK. However, it should be noted that the rules given in EC2 are specific to mesh fabrics made of high bond wires, whereas the rules in BS 8110 permit the use of plain round wires. Hence the results of the two codes are not directly compatible.

EC2 recommends that laps be made in zones where the effect of the rare combination of loads is not more than 80 per cent of the design strength of the section (clause 5.2.4.2.1). Where this condition is not fulfilled the effective depth of the steel should be calculated on the layer with the least (value) effective depth.

The permissible percentage of the main reinforcement which may be lapped in any one section is 100 per cent of the total steel cross area if $A_s/s \leq$ 1200 mm^2/m. This should not exceed 60 per cent if $A_s/s >$ 1200 mm^2/m and if the wire mesh is an interior mesh. 'Interior' assumes that two layers of mesh exist in that face, e.g. bottom reinforcement.

Joints of multilayered mesh should be staggered at 1.3 × lap length. The lap length, l_s, is calculated from

$$l_s = \alpha_2 l_b \frac{A_{s,req}}{A_{s,prov}} \not< l_{s,min} \qquad\qquad [6.3]$$

where

$$\alpha_s = 0.4 + \frac{A_s/s}{800} \not< 1 \quad \text{or} \quad \not> 2$$

$$l_{s,min} = 0.3\alpha_2 l_b \not< 200\,\text{mm} \quad \text{or} \quad \not> s_t$$

where s_t is the spacing of transverse wires.

BS 8110 allows a higher bond stress for fabric, but only if the number of intersections in anchorage length is at least equal to $4A_{s,req}/A_{s,prov}$. This limitation often prevents the use of the higher bond stress (clause 3.12.8.5).

EC2 provides less onerous rules for the laps of transverse distribution reinforcement (clause 5.2.4.2.2). It may all be lapped in one direction and normally requires only the length of one mesh to be lapped.

6.7   Anchorage of Links and Shear Reinforcement (clause 5.2.5)

EC2 accepts that anchorage of links with high bond bars is effective with 90° bends (clause 5.2.5). However, it normally expects such reinforcement to be hooked or welded to transverse reinforcement (longitudinal to the section). A longitudinal bar is required inside a hook or bend (see Fig. 5.7 of EC2).

The anchorage is considered satisfactory if for a hook or bend of 135° or more the length the straight bar extends beyond the bar by 4ϕ (NAD) or 50 mm whichever is greater. For 90° bends the straight bar should extend 8ϕ (NAD) or 70 mm whichever is greater (see Fig. 5.7 of EC2).

If welded bars are used to anchor the link, these can be by two bars near the end. Their size should not be less than $0.7 \times$ the diameter of the link bar. Their distance apart should not be less than $2 \times$ the diameter of the link bar or 20 mm, or more than 50 mm. A single welded transverse bar can be used provided its diameter is not less than $1.4 \times$ the diameter of the link.

BS 8110 requires 4ϕ of straight bar beyond the bend of a hook and 8ϕ beyond a 90° bend (clause 3.12.8.6). It also allows the use of welded bars in a similar way to EC2 (clause 3.12.8.7).

6.8   Additional Rules for High Bond Bars Exceeding 32 mm Diameter (clause 5.2.6)

EC2 requires some specific rules when bars of larger diameter than 32 mm are used (clause 5.2.6). Such bars should only be used in elements with depth not less than 15ϕ. Adequate crack control should be ensured either by using surface reinforcement or by calculation. The bond stress of such bars should be reduced by a factor of $(132 - \phi)/100$. The anchorage of such large bars should be as for straight bars of normal diameter or by mechanical devices. They should not be anchored in tension zones.

Additional transverse reinforcement is required in the direction parallel to the lower face

$$A_{st} = n_1 \times 0.25A_s$$

and in the direction perpendicular to the lower face

$$A_{sv} = n_2 \times 0.25A_s$$

where A_s is the cross-sectional area of the anchored bar, n_1 the number of layers with bar anchored at the same point in the member and n_2 the number of bars anchored in each layer. Figure 5.8 of EC2 shows how these rules should be applied.

If surface reinforcement is used for crack control, the area should not be less than $0.01A_{ct.ext}$ in the direct perpendicular to the large-diameter bars and $0.02A_{ct.ext}$ parallel to those bars as shown in Fig. 5.15 of EC2.

These rules will come as a shock to those who are accustomed to designing to BS 8110 using large-diameter bars (40 or 50 mm). The most affected structures will be large concrete rafts where bars have to be lapped under full tension. EC2 does not permit this.

6.9   Bundled High Bond Bars (clause 6.2.7)

EC2 is a little more restrictive than BS 8110. It allows bundles of up to four bars in compression zones including laps and three bars in other situations (clause 5.2.7). All bars in a bundle must be of the same diameter and of the same type and grade. For design the bundle is replaced by a notional bar having the same sectional area and the same centre of gravity as the bundle. The equivalent diameter should not exceed 55 mm. This is about the value equivalent to three 32 mm diameter bars.

The spacing rules for bundled bars in EC2 depend on the equivalent diameter of the bar except that the clear distance should be considered from the actual external contour of the bundles of bars. The concrete cover measured from the actual external contour of the bundles should be greater than the equivalent diameter.

BS 8110 allows up to four bars in a bundle for all situations. The equivalent diameter is calculated in the same way as for EC2 and this is used for both spacing and cover calculations (i.e. is less conservative then EC2).

6.9.1   Anchorage and Joints of Bundled Bars (clause 5.2.7.2)

EC2 only allows straight bar anchorage or lapping. They should be staggered.

The staggers for bundles of two, three or four bars should be 1.2, 1.3 and 1.4 times the anchorage length of the individual bars respectively. The bars should be lapped one by one.

BS 8110 limits the number of bundled bars at laps to four, thus the rules for bundled bars to EC2 are a little more conservative than those for BS 8110.

6.10   Prestressing Units (clause 5.3)

EC2 requires that pretensioned tendons be spaced apart. In post-tensioned members bundled ducts are not normally permitted except if a pair is placed with one vertically above the other. BS 8110 requires sufficient horizontal space between ducts to allow a vibrator to be inserted.

6.10.1 Concrete Cover (clause 5.3.2)

EC2 refers to clause 4.1.3.3 of the design section. For pretensioned members the minimum cover should not be less than twice the diameter of the tendon and for ribbed wires three times the diameter of the wire. For post-tensioned members the minimum cover to the duct should not be less than the diameter of the duct. For rectangular ducts the cover should not be less than the lesser dimension of the duct cross-section or half the greater dimension. BS 8110 refers to the rules for reinforcement, and in addition requires that the cover to ducts should be $\not< 50$ mm (clause 4.12.3.2).

6.10.2 Horizontal and Vertical Spacing (clause 5.3.3)

EC2 gives specific rules for pre- and post-tensioning.

6.10.2.1 Pre-tensioning
The minimum clear horizontal and vertical spacing of individual tendons is given in Fig. 6.3

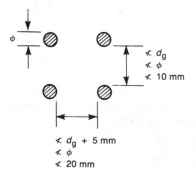

Figure 6.3 Minimum clear spacing for pretensioned tendons

6.10.2.2 *Post-tensioning*

Except for paired ducts, the minimum clear spacing between individual ducts should be as follows:

Horizontal: $\not< \phi_{\text{duct}}$ or 40 mm
Vertical: $\not< \phi_{\text{duct}}$ or 50 mm

where ϕ_{duct} denotes the diameter of the duct.

BS 8110 refers to the rules for reinforcement (clause 4.12.4.2). It also requires sufficient space between bonded wires or strands at ends of members to allow the transmission length to be developed. Where two or more groups or wires or strands are widely spaced the possibility of longitudinal splitting should be considered.

Also in BS 8110, the clear distance between ducts or between ducts and other tendons should not be less than the following, whichever is the greatest:

1. $h_{\text{agg}} + 5$ mm, where h_{agg} is the nominal maximum size of the coarse aggregate;
2. In the vertical direction, the vertical internal dimension of the duct;
3. In the horiztonal direction, the horizontal internal dimension of the duct; where internal vibrators are used sufficient space should be provided between ducts to enable the vibrator to be inserted.

Where two or more rows of ducts are used, the horizontal gaps between the ducts should be vertically in line wherever possible, for ease of construction. BS 8110 gives specific rules for curved tendons and provides tables for cover and spacing of ducts (Tables 4.10 and 4.11 of BS 8110).

6.10.3 Anchorages and Couplers for Prestressing Tendons (clause 5.3.4)

EC2 requires that the full strength of tendons should be developed taking account of any repeated, rapidly changing action effects. BS 8110 requires that the overall equilibrium of the end block should be checked and spalling of the concrete from the loaded face around the anchorages be prevented (clause 4.11.1).

EC2 provides specific rules for couplers not included in BS 8110. In general, couplers should be located away from intermediate supports. The use of couplers on more than 50 per cent of the tendons of one cross-section should be avoided.

6.11 Columns (clause 5.4.1)

6.11.1 Longitudinal Reinforcement (clause 5.4.1.2)

EC2 requires that the diameter of the longitudinal bars should be $\not< 12$ mm. The minimum amount of longitudinal reinforcement $A_{\text{s,min}}$ is governed by

$$A_{\text{s,min}} = \frac{0.15N_{\text{sd}}}{f_{\text{yd}}} \not< 0.003A_{\text{c}} \qquad [6.4]$$

where f_{yd} is the design yield strength of the reinforcement, N_{sd} the design axial compression force and A_c the cross-section of the concrete. Even at laps, the area of reinforcement should not exceed the upper limit $0.08A_c$. The longitudinal bars should be distributed around the periphery of the section. For columns which have a polygonal cross-section, at least one bar shall be placed at each corner. For columns of circular cross-section the minimum number of bars is six.

BS 8110 requires that the minimum percentage of compression reinforcement should not be less than 4 (clause 3.12.5). The maximum percentage should not exceed 6 per cent for vertically cast columns and 8 per cent for horizontally cast columns. At laps it should not exceed 10 per cent (clause 3.12.6). Where reinforcement is not required to resist the load the column is treated in a similar manner to plain walls (clause 3.9.4).

6.11.2 Transverse Reinforcement (clause 5.4.1.2.2)

Both EC2 and BS 8110 require that the diameter should be $\not< 6$ mm or one quarter of the maximum diameter of the longitudinal bars. Both codes require the spacing of links to be $\not> 12$ times the minimum diameter of longitudinal bars and EC2 also requires that this should be $\not> 300$ mm. EC2 requires that the spacing be reduced by a factor of 0.67:

1. In sections located above and below a beam or slab over a height equal to the larger dimension of the column cross-section;
2. Near lapped joints; if the maximum diameter of the longitudinal bars is greater than 20 mm (NAD).

Where the direction of the longitudinal bars changes (e.g. at changes in column size) the spacing of transverse reinforcement should be calculated, while taking account of the lateral forces involved.

BS 8110 requires that, at laps where both bars exceed 20 mm diameter, the spacing of links be reduced to not greater than 200 mm (clause 3.12.8.12). EC2 requires that every longitudinal bar (or group of longitudinal bars) placed in a corner should be held by transverse reinforcement. A maximum of five bars in or close to each corner can be secured against buckling by any one set of transverse reinforcement. BS 8110 requires that every corner bar and each alternate bar (or pair or bundle) in an outer layer of reinforcement should be supported by a link passing round the bar and having an included angle of not more than 135°. No bar within the compression zone should be further than 150 mm from a restrained bar (clause 3.12.7.2).

6.12   Beams (clause 5.4.2)

6.12.1 Minimum Area of Tension Reinforcement (clause 5.4.2.1.1)

EC2 sets a minimum area of tensile reinforcement based on f_{yk}:

$$A_{st,min} \geq 0.6b_t d/f_{yk} \geq 0.0015b_t d \qquad [6.5]$$

where b_t is the width tension zone, d the effective depth and f_{yk} the characteristic yield strength. EC2 also refers to clause 4.4.2 — minimum reinforcement area to control cracking.

The minimum values in BS 8110, Table 3.27, are based on the full depth, h, not d as for EC2. For high yield reinforcement ($f_y = 460$). Here $A_{st,min} \geq 0.0013 b_w h$ corresponds to the value in EC2 for $d = 0.87h$, i.e. the two codes give similar limitations.

6.12.2 Maximum Area of Tension and Compression Reinforcement

EC2 sets a limit to the areas of tension and compression areas of reinforcement. They should be $\not> 0.04 A_c$. This is taken to mean that neither should exceed this limit. BS 8110 gives the same limits.

6.12.2.1 Monolithic construction
EC2 imposes two rules which do not exist in BS 8110

1. Even when simple supports have been assumed in design, the section should be designed for bending moment arising from partial fixity of at least 25 per cent of the maximum bending moment in the span.
2. At intermediate supports of continuous beams, the total amount of tensile reinforcement, A_s, of a flanged cross-section may be divided approximately equally between the internal and external parts of the flange (see Fig. 6.4).

6.12.3 Curtailment of Longitudinal Reinforcement (clause 5.4.2.1.3)

As explained under section 6.5, EC2 increases the curtailment length of reinforcement compared to BS 8110 by the value a_1. The value of a_1 depends on how the shear reinforcement is calculated. If it is calculated according to the standard method (see clause 4.3.2.4.3 of EC2)

$$a_1 = z(1 - \cot \alpha)/2 \geq 0$$

α being the angle of the shear reinforcement with the longitudinal axis. If the shear reinforcement is calculated according to the variable strut inclination method (see clause 4.3.2.4.4 of EC2),

$$a_1 = z(\cot \theta - \cot \alpha)/2 \geq 0$$

θ being the angle of the concrete struts with the longitudinal axis. Normally z can be taken as $0.9d$. If links are used $\alpha = 90°$, hence for the standard method $a_1 = 0.5Z$. This can be approximated to $0.5d$. For reinforcement in the flange, placed outside the web, a_1 should be increased by the distance of the bar from the web (distance x in Fig. 6.4). Both EC2 and BS 8110 require a minimum anchorage length of d beyond the point that the bar is required.

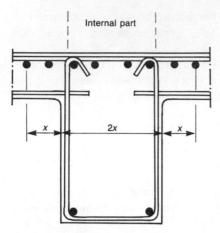

Figure 6.4 Internal and external parts of a T-beam

The effect of these rules is shown in Fig. 6.5, where EC2 requires up to 50 per cent more curtailment length than BS 8110.

6.12.3.1 Bent-up Bars

EC2 requires the anchorage length of bent-up bars, which contribute to shear resistance should not be less than $1.3l_{b,net}$ in the tension zone and $0.7l_{b,net}$ in the compression zone. No equivalent rules exist in BS 8110.

6.12.4 Anchorage Reinforcement at an End Support (clause 5.4.2.1.4)

EC2 requires that at least 50 per cent of the area reinforcement required in the span should continue into the support. This is the same as the simplified rules

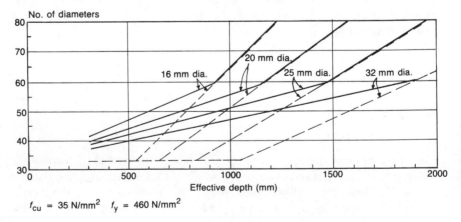

$f_{cu} = 35 \text{ N/mm}^2$ $f_y = 460 \text{ N/mm}^2$

Figure 6.5 Comparison of curtailment rules for beams. (——) EC2; (---) BS 8110

of BS 8110 (Fig. 3.24 of BS 8110). EC2 also demands that the reinforcement anchored at the support should be capable of resisting a force

$$F_s = V_{sd}a_1/d + N_{sd} \qquad\qquad [6.6]$$

where V_{sd} is the design shear force, N_{sd} the design axial tensile force, if present and a_1/d are as defined previously. There is no equivalent to this in BS 8110.

The anchorage length, according to EC2, is measured from the line of contact between the beam and its support. It should be taken as follows:

> For a direct support: $\frac{2}{3}l_{b,net}$ (see Fig. 6.6a)
> For an indirect support: $l_{b,net}$ (see Fig. 6.6b)

where $l_{b,net}$ is calculated according to Eq. 6.2. These rules originate from DIN 1045 and the meaning is assumed to be the same. A 'direct' support should be assumed if the beam (or slab) is supported by a wall, column or beam whose depth is at least twice the depth of the supported beam (or slab). Otherwise it should be considered as an 'indirect' support.

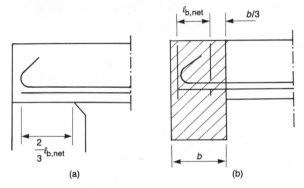

Figure 6.6 Anchorages of bottom reinforcement at end supports. (a) Direct support; (b) indirect support

BS 8110 has very different clauses to determine the anchorage of such bars. Each tension bar should be anchored by one of the following:

1. An effective anchorage length equivalent to 12 times the bar size beyond the centre line of the support; no bend or hook should begin before the centre of the support.
2. An effective anchorage length equivalent to 12 times the bar size plus $d/2$ from face of the support, where d is the effective depth of member; no bend or hook should begin before $d/2$ from the face of the support.

6.12.5 Moment Connection Between Beams and Edge Columns (or Walls)

There are no specific rules given in either code. However, the methods for each can be deduced from the anchorage rules. Typically a calculation is required to determine how far the top bars from the beam should extend down the column.

The procedure using EC2 is as follows:

1. Determine the length of curtailment that is required from the face of the support, $l_{b,net}$. Here α_a will be 0.7 if there is a 90° bend (clause 5.2.3.4.1). Since the support is 'direct', $l_{b,net}$ can be reduced by a factor of $\frac{2}{3}$ (clause 5.4.2.1.4(3)). If bond conditions are 'poor' (clause 5.2.2.1) then $l_{b,net}$ should be increased by a factor of $1/0.7$ (clause 5.2.2.2(2)).
2. If the effective value of $l_{b,net}$ in (1) above exceeds the distance available in the column, the bars should extend down further than that for a standard bend. The required length is calculated as the sum of the horizontal and vertical components of the anchorage without taking any enhancement for the bend conditions in the column which are always considered 'good' (clause 5.2.2.1).
3. If the required length of bar extends down the column beyond the soffit of the beam it may be reasonable to provide a further bend (U-bar within the depth of the beam). If a standard bend is to be used at the bottom of the vertical leg it may be factored by α_a equal to 0.7. If the horizontal length of bar after this bend is required to be more than 4ϕ (NAD), then the same procedure as for (2) above should be adopted.

The procedure using BS 8110 is simpler. The effective length of bar can be calculated taking account of one or two bends, if necessary (clause 3.12.8.23).

Example 6.1

Assume $f_{ck} = 30(f_{cu} = 37)$, $\phi = 25$, $A_{s,req}/A_{s,prov} = 1$.

Standard Hook

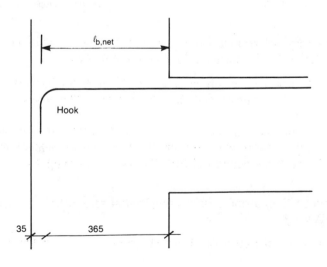

EC2

From Table 5.3 of EC2 $f_b =$ 3 N/mm^2. From Eq. 6.5

$$l_{b,net} = \alpha_a l_b \frac{A_{s,req}}{A_{s,prov}}$$

$$\alpha_a = 0.7 \quad \frac{A_{s,req}}{A_{s,prov}} = 1$$

$$l_b = \frac{25}{4} \times \frac{0.87 \times 460}{3}$$

$$= 834 \text{ mm}$$

Direct support, but 'poor' bond, hence

$$l_{b,net} = 0.7 \times \frac{2}{3} \times \frac{834 \text{ mm}}{0.7}$$

$$= 556 \text{ mm} > 365 \text{ mm}$$

Hence vertical leg required.

BS 8110

From Table 3.28 and Eq. 49 of BS 8110

$$f_b = 0.5\sqrt{37} = 3.04 \text{ N/mm}^2$$

Anchorage length

$$l = \frac{F_s}{\pi \phi_s f_b}$$

$$= \frac{0.87 f_y \times \pi \phi_s^2}{4\pi \phi_s f_b}$$

$$= \frac{0.87 \times 460 \times 25}{4 \times 3.04} = 823 \text{ mm}$$

Check to clause 3.12.8.23:

1. Internal radius of bend

    $$r_{int} = 4 \times 25 = 100 \text{ mm}$$

    $$4 \times r_{int} = 400 \text{ mm} \quad \text{or}$$

    $$12 \times \phi = 300$$

    whichever is less. Or
2. Actual length of bend is

    $$\frac{4.5 \times 35 \times \pi}{2} = 176.7 \text{ mm}$$

Choose the greater between (1) and (2) which is 300 mm.

$$l_{h,eff} = 240 + 300 \text{ mm}$$
$$= 540 \text{ mm} < 823 \text{ mm}$$

Hence vertical leg required.

Vertical Leg

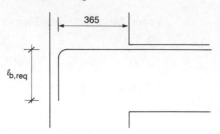

EC2   BS 8110

Anchorage length = horizontal $l_{b,req} = 823 - 240 - 300 + 100 + 125$
anchorage (poor bond: 0.7 $= 508$ mm
transverse pressure: $\frac{2}{3}$) +
vertical anchorage (good bond)
without $\alpha_a = 0.7$ for bend.

$l_{b,req} = 834 - 365 \times 0.7/\frac{2}{3} = 451$ mm

U-bar with Standard 90° Bend at Bottom

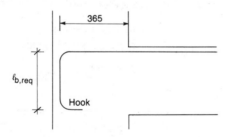

EC2   BS 8110

α_a = 0.7 for bend on vertical leg   Assuming $l_{b,req} = 350$ mm
$l_{b,req}$ = $(834 - 365 \times 0.7/\frac{2}{3})0.7$ $l_{b,net} = 240 + 300 + 300$
         = $451 \times 0.7 = 316$ mm $= 840$ mm > 823 mm

It should be noted, however, that the
minimum value permitted with two
bends is $14\phi = 350$ mm (see
BS 4466 : 1989 clause 7.10). Hence
the equivalent

$\quad l_{b,net} = 834 + 34/0.7$
$\qquad\quad = 883$ mm > 834 mm

6.12.6 Anchorage of Bottom Reinforcement at Intermediate Supports (clause 5.4.2.1.5)

EC2 requires that the span bottom reinforcement should pass into the support by at least 10ϕ or not less than the diameter of the mandrel for a hook or bend (see Fig. 6.7). It is also recommended that the reinforcement is continuous and able to resist accidental positive moments (settlement of the support, explosion, etc.).

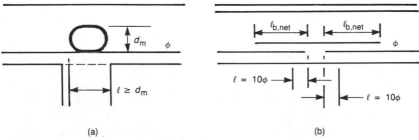

| (a) | (b) |

Figure 6.7 Anchorage at intermediate supports (d_m = diameter of mandrel)

BS 8110 does not contain such rules. However, the simplified curtailment rules given in Fig. 3.24 recommend that at least 30 per cent of the area of reinforcement provided in this span should extend at least $d/2$ into the support. This is often detailed as shown in Fig. 6.8.

The EC2 rules prevent the use of preformed cages as is common in the UK, since the bottom bars are required to extend into the support in addition to lapping with a support bar.

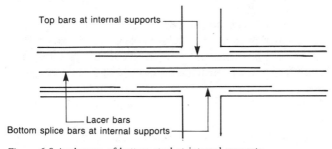

Figure 6.8 Anchorage of bottom steel at internal supports

6.12.7 Shear Reinforcement (clause 5.4.2.2)

Both codes require that the shear reinforcement should be placed at an angle of between 45 and 90° with the mid-plane of the structural element. Both codes require that at least 50 per cent of the necessary shear reinforcement is provided by links.

EC2 allows combinations of the following:

1. Links enclosing the longitudinal tensile reinforcement and the compression zone;
2. Bent-up bars;
3. Shear assemblies in the form of cages, ladders, etc. of high bond bars which are cast in without enclosing the longitudinal reinforcement (see Fig. 6.9), but should be properly anchored in the compression and tension zones. There is no equivalent to this clause in BS 8110.

A lap joint on the leg of link near the surface of a web is allowed only for high bond bars. This rule does not exist in BS 8110 and laps may occur in the vertical leg for both smooth and high bond bars.

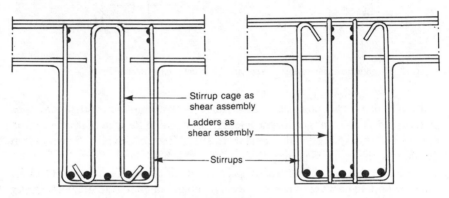

Figure 6.9 Examples for combinations of links and shear reinforcement

6.12.7.1 *Minimum Shear Reinforcement (clause 5.4.2.2(5))*

EC2 provides Table 5.5 in which the minimum value of shear reinforcement ratio $A_{sw}/(sb_w \sin \alpha)$ depends on the concrete and reinforcement strength. BS 8110 requires that minimum links should provide a design shear resistance of $0.4 \, \text{N/mm}^2$. This compares well up to $f_{ck} = 35 \, \text{N/mm}^2$. EC2 limits the diameter of shear reinforcement for plain round bars to $\not> 12 \, \text{mm}$ (clause 5.4.2.2(6)).

6.12.7.2 *Maximum Longitudinal Spacing of Shear Reinforcement*
(clause 5.4.2.2(7))

EC2 relates the maximum spacing of links with the ratio of the applied shear force V_{sd} to the maximum resistance, V_{rd2}. If

$$V_{sd} \leq \tfrac{1}{5}V_{rd2} \qquad s_{max} = 0.8d \not> 300 \, \text{mm}$$

if

$$\tfrac{1}{5}V_{rd2} < V_{sd} \leq \tfrac{1}{5}V_{rd2} \qquad s_{max} = 0.6d \not> 300 \, \text{mm}$$

and if

$$V_{sd} > \tfrac{2}{3}V_{rd2} \qquad s_{max} = 0.3d \not> 200 \, \text{mm}$$

BS 8110 requires only that links should not exceed a spacing in the direction of the span of $0.75d$. Although this appears very different for situations of high shear, it is unlikely to lead to much practical difference since the amount of shear reinforcement required in such situations will also require the pitch to be small.

6.12.7.3 Maximum Longitudinal Spacing of Bent-up Bars (clause 5.4.2.2(8))
EC2 requires that

$$s_{\max} = 0.6d(1 + \cot \alpha) \tag{6.7}$$

where α is the angle between the shear reinforcement and the main steel.

BS 8110 considers a series of bent-up bars and limits the effective spacing of the compression struts to $1.5d$ (clause 3.4.5.6) (see Fig. 6.10).

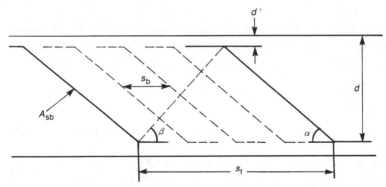

Figure 6.10 System of bent-up bars

6.12.7.4 Transverse Spacing of Legs of Links
EC2 uses similar rules to those for longitudinal spacing. If

$$V_{sd} \le \tfrac{1}{5}V_{rd2} \qquad s_{\max} = d \text{ or } 800\,\text{mm whichever is less}$$

where $V_{sd} > \tfrac{1}{5}V_{rd2}$ Eqs 5.18 or 5.19 applies.

BS 8110 restricts the transverse spacing of link legs such that no longitudinal tension bar is more than 150 mm from a vertical leg: this spacing should not in any case exceed d.

6.12.8 Torsional Reinforcement (clause 5.4.2.3)

EC2 provides the following rules:

1. The torsion links should be closed and be anchored by means of laps or according to (a) of Fig. 6.11 and form an angle of 90° with the axis of the structural element.
2. The provisions of clause 5.4.2.2(3)–(6), which apply to shear reinforcement, are also valid for the longitudinal bars and links of beams subjected to torsion.

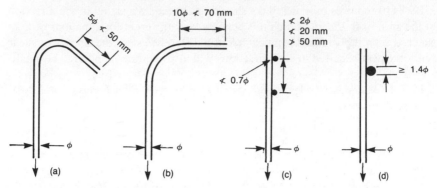

Figure 6.11 Anchorage of links

3. The longitudinal spacing of the torsion links should not exceed $u_k/8$, where u_k is defined in clause 4.3.3.1. For a rectangular section (h, b), this spacing will be equal to

$$\frac{h^2 + bh + b^2}{4(h + b)}$$

4. The spacing in (3) above should also satisfy the requirements in clause 5.4.2.2(7) for maximum spacing of links.
5. The longitudinal bars should be so arranged that there is at least one bar at each corner, the others being distributed uniformly around the inner periphery of the links, spaced at no more than 350 mm centres.

BS 8110 provides the following rules. Torsion reinforcement should consist of rectangular closed links together with longitudinal reinforcement. This reinforcement is additional to any requirements for shear or bending and should be such that

$$\frac{A_{sv}}{s_v} > \frac{T}{0.8x_1 y_1 (0.87 f_{yv})}$$

$$A_s > \frac{A_{sv} f_{yv} (x_1 + y_1)}{s_v f_y}$$

Note: f_y and f_{yv} should not be taken as $> 460\ \text{N/mm}^2$.

The spacing s_v should not exceed the least of x_1, $y_1/2$ or 200 mm. The links should be of the closed type complying with shape code 74 of BS 4466. Longitudinal torsion reinforcement should be distributed evenly around the inside perimeter of the links, the clear distance between these bars should not exceed 300 mm and at least four bars, one in each corner of the links, should be used. Additional longitudinal reinforcement required at the level of the tension or compression reinforcement may be provided by using larger bars than those required for bending along. The torsion reinforcement should extend a distance at least equal to the largest dimension of the section beyond where it theoretically ceases to be required.

In the component rectangles the reinforcement cages should be detailed so that they interlock and tie the component rectangles of the section together. Where the torsional shear stress in a minor component rectangle does not exceed $v_{t,min}$ no torsion reinforcement need be provided in that rectangle.

6.12.9 Surface Reinforcement (clause 5.4.2.4)

EC2 provides specific rules for the use of surface reinforcement for which there is no equivalent in BS 8110. In clause 4.1.7 of Part 2, BS 8110 states that 'welded steel fabric as supplementary reinforcement is sometimes used to prevent spalling; it is then placed within the cover at 20 mm from the concrete face. There are practical difficulties in keeping the fabric in place and in compacting the concrete; in certain circumstances there would also be a conflict with durability recommendations of this standard'. This recognizes the real difficulties of achieving an acceptable final product.

EC2 reflects German practice and provides specific rules:

1. In certain cases it may be necessary to provide surface reinforcement either to control cracking or to ensure adequate resistance to spalling of the cover.
2. Skin reinforcement to control cracking should normally be provided in beams over 1 m deep (see clause 4.4.2.3(4) of EC2).
3. Surface reinforcement to resist spalling arising, for example, from fire or where bundled bars or bars > 32 mm diameter are used, should consist of wire mesh or small diameter high bond bars and be placed outside the links as indicated in Fig. 5.15 of EC2.
4. The minimum cover needed for the surface reinforcement is given in clauses 4.1.3.3(6) and (7).
5. The area of surface reinforcement $A_{s,surf}$ should be not less than $0.01A_{ct,ext}$ in the direction parallel to the beam tension reinforcement. Here $A_{ct.ext}$ denotes the area of the tensile concrete external to the links, defined by Fig. 5.15 of EC2.
6. The longitudinal bars of the surface reinforcement may be taken into account as longitudinal bending reinforcement and the transverse bars as shear reinforcement provided that they meet the requirements for the arrangement and anchorage of these types of reinforcement.
7. Any surface reinforcement in prestressed beams can be taken into account as in (5) and (6) above.

6.13   Cast *in situ* Solid Slabs (clause 5.4.3)
6.13.1 Flexural Reinforcement (clause 5.4.3.2)

EC2 requires an increase of curtailment over BS 8110 of a_1 similar to that for beams. For slabs the value of a_1 should be taken as d (clause 5.4.3.1(1)). Cut-off bars should then be not less than d beyond this point (clause 5.4.2.1.3(2)).

The rules for bottom reinforcement at end and intermediate supports are the same as those for beams (clauses 5.4.2.1.4(1)–(3), 5.4.2.1.5(1)–(2)). Secondary transverse reinforcement is required in one-way slabs of at least 20 per cent of the principal reinforcement (clause 5.4.3.2.1(2)). Otherwise the minimum and maximum steel percentages are as for beams. The maximum spacing of bars is as follows (clause 5.4.3.2.1(4)):

Principal and secondary reinforcement: $3h \not> 500$ mm   (NAD)

BS 8110, Fig. 3.25, provides simplified rules for curtailment of reinforcement. These can be used where

1.   The slabs are designed for predominantly uniformly distributed loads;
2.   In the case of continuous slabs, the design has been carried out for the single load case of maximum design load on all spans and the spans are approximately equal.

The maximum spacing of bars to BS 8110 is similar to that for EC2, but includes some additional rules (clause 3.12.11.27). In BS 8110 unless crack widths are checked by direct calculation, the following rules will ensure adequate control of cracking for slabs subjected to normal internal and external environments:

1.   No further check is required on bar spacing if either:
     (a)   grade 250 steel is used and the slab depth does not exceed 250 mm; or
     (b)   grade 460 steel is used and the slab depth does not exceed 200 mm; or
     (c)   the reinforcement percentage ($100A_s/bd$) is < 0.3 where A_s is the minimum recommended area, b the breadth of section at the point considered and d the effective depth.
2.   Where none of conditions (a), (b) and (c) apply, the bar spacing should be limited to the values given in Table 3.30 of BS 8110 for slabs where the reinforcement exceeds 1 per cent, or the values given in Table 3.30 divided by the reinforcement percentage for lesser amounts.

6.13.2   Reinforcement in Slabs Near Supports (clause 5.4.3.2)

Where reinforcement is needed to distribute cracking arising from shrinkage and temperature effects, BS 8110 refers to the rules for plain walls (clause 3.9.4.19 and 20).

EC2 requires that at least half the calculated span reinforcement should continue up to the support and be anchored therein. The simplified rules of BS 8110 only require that 40 per cent of the span reinforcement extend into the support; 100 per cent of the span reinforcement must extend to within $0.2l$ for continuous slabs and $0.1l$ for simple supported slabs.

EC2 requires that where partial fixity occurs along one side of a slab but has not been taken account of in the analysis, top reinforcement should be provided equivalent to at least a quarter of the maximum moment in the adjacent span.

The wording is unclear but it is assumed that this reinforcement should extend to not less than $0.2l$.

BS 8110 also requires such reinforcement at end supports where simple support has been assumed in the assessment of moments (clause 3.12.10.3.1). An amount of reinforcement equal to half the area of bottom steel at mid-span, but not less than the minimum value, should be provided in the top of the slab at the support. It should have a full effective tensile anchorage into the support and extend ≮ $0.15l$ or 45 times the bar size into the span. Bottom reinforcement may be detailed as follows:

1. As indicated in Fig. 3.25 for a simply supported end, in which case the shear strength at the support may be based on the area of bottom steel continuing into the support; or
2. As indicated in Fig. 3.25 for a simply supported end except that the bottom steel is stopped at the line of effective support; in this case the shear strength at the support should be based on the area of top steel.

6.13.3 Corner Reinforcement (clause 5.4.3.2.3)

EC2 notes that if the lifting of a corner of a slab is restrained, suitable reinforcement should be provided. BS 8110 provides more detailed rules for such situations (clauses 3.5.3.2(5) (6) and (7)).

6.13.4 Reinforcement at Free Edges (clause 5.4.3.2.4)

EC2 requires that:

1. Along a free (unsupported) edge, a slab should normally contain longitudinal and transverse reinforcement generally arranged as shown in Fig. 6.12.
2. The normal reinforcement provided for a slab may act as edge reinforcement.

BS 8110 does not provide any equivalent recommendations. However, for flat slabs, minimum top reinforcement is required to extend into the span at least $0.1l$ or an anchorage length whichever is greater.

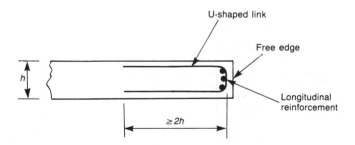

Figure 6.12 Edge reinforcement for a slab

6.13.5 Shear Reinforcement

Both EC2 and BS 8110 restrict the use of shear reinforcement to slabs with a depth of at least 200 mm. EC2 allows bent-up bars or shear assemblies to provide all the required shear reinforcement for $V_{sd} \leq \frac{1}{3}V_{rd2}$ (clause 5.4.3.3(3)). For greater values of V_{sd} only 50 per cent of resistance may be provided by bent-up bars. This is in accordance with clause 5.4.2.2 which also applies to slabs.

Other detailed rules for EC2 are as follows (clauses 5.4.3.3.(4)−(7)):

1.   The maximum longitudinal spacing of successive series of links is given by Eqs 5.17−5.19 of EC2, neglecting the limits given in millimetres. The maximum longitudinal spacing of bent-up bars is $s_{max} = d$. In practice the minimum spacing limit may be reduced to $0.3d$ (see 4.4.4), which is very conservative compared with BS 8110.
2.   The distance between the inner face of a support, or the circumference of a loaded area, and the nearest shear reinforcement taken into account in the design should not exceed $d/2$ for bent-up bars. This distance should be taken at the level of the flexural reinforcement; if only a single line of bent-up bars is provided, their slope may be reduced to 30° (Fig. 6.13b).
3.   It may be assumed that one bent-up bar takes up the shear force over a length of $2d$.
4.   Only the following reinforcement may be taken into account as punching shear reinforcement:
     (a) reinforcement located in a zone bounded by a contour line situated at a distance not exceeding $1.5d$ or 800 mm, whichever is the smaller, from the periphery of the loaded area; this condition applies in all directions;

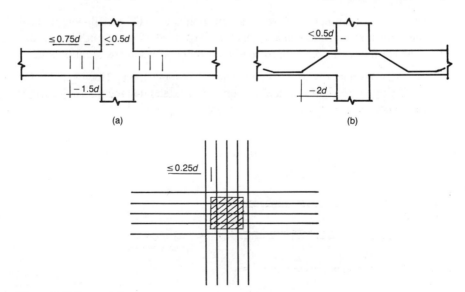

(a)   (b)

Figure 6.13 Shear reinforcement near support

(b) bent-up bars passing over the loaded area (Fig. 6.13b) or at a distance not exceeding $d/4$ from the periphery of this area (Fig. 6.13c).

5.   Further critical perimeters should be considered outside the area reinforced for shear (clause 4.3.4.5.2(3)).

BS 8110 has special rules for flat slab as follows. The shear capacity is checked first on a perimeter $1.5d$ from the face of the loaded area. If the calculated shear stress does not exceed v_c then no further checks are needed.

If shear reinforcement is required, then it should be provided on at least two perimeters within the zone indicated in Fig. 6.14. The first perimeter of reinforcement should be located at approximately $0.5d$ from the face of the loaded area and should contain not less than 40 per cent of the calculated area of reinforcement.

The spacing of perimeters of reinforcement should not exceed $0.75d$ and the spacing of the shear reinforcement around any perimeter should not exceed $1.5d$. The shear reinforcement should be anchored round at least one layer of tension reinforcement. The shear stress should then be checked on successive perimeters

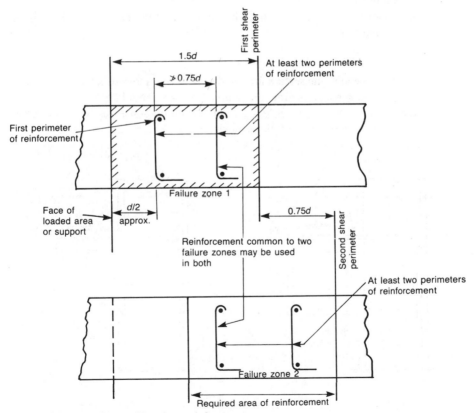

Figure 6.14 Zones for punching shear reinforcement

at $0.75d$ intervals until a perimeter is reached which does not require shear reinforcement.

In providing reinforcement for the shear calculated on the second and subsequent perimeters, that provided for the shear on previous perimeters and which lies within the zone shown in Fig. 6.14 should be taken into account.

6.14   Corbels (clause 5.4.4)

Both EC2 and BS 8110 use strut and tie models to determine the detailing rules.

6.14.1   EC2

1. The reinforcement, corresponding to the ties considered in the design model (clause 2.5.3.7), should be fully anchored beyond the node under the bearing plate by using U-hoops or anchorage devices, unless a length $l_{b,net}$ is available between the node and the front of the corbel. Here, $l_{b,net}$ should be measured from the point where the compression stresses change their direction.
2. In corbels with $h_c \geq 300$ mm, when the area of the primary horizontal tie A_s is such that

$$A_s \geq 0.4A_c f_{cd}/f_{yd} \qquad\qquad [6.8]$$

where A_c denotes the section area of the concrete in the corbel at the column, then closed stirrups, having a total area $\not< 0.4A_s$, should be distributed over the effective depth d in order to cater for splitting stresses in the concrete strut. They can be placed either horizontally (Fig. 6.15a) or inclined (Fig. 6.15b).

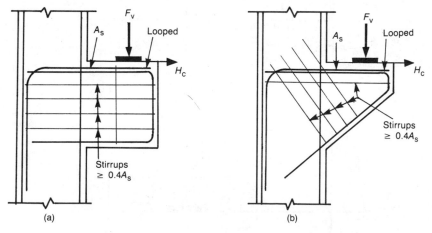

Figure 6.15 (a) Reinforcement of a corbel with horizontal stirrups; (b) reinforcement of a corbel with inclined stirrups

6.14.2 BS 8110 (clause 5.2.7.2.2 to 4)

At the front face of the corbel, the reinforcement should be anchored either by

1. Welding to a transverse bar of equal strength; in this case the bearing area of the load should stop short of the face of the support by a distance equal to the cover of the tie reinforcement; or
2. By bending back the bars to form a loop; in this case the bearing area of the load should not project beyond the straight portion of the bars forming the main tension reinforcement.

6.14.2.1 Shear Reinforcement

Shear reinforcement should be provided in the form of horizontal links distributed in the upper two-thirds of the effective depth of root of the corbel; this reinforcement should be not less than one-half of the area of the main tension reinforcement and should be adequately anchored.

6.14.2.2 Resistance to Applied Horizontal Force

Additional reinforcement connected to the supported member in accordance with clause 5.3 of BS 8110 should be provided to transmit this force in its entirety.

6.15   Deep Beams (clause 5.4.5)

BS 8110 does not cover the design and detailing of deep beams. EC2 provides the following rules.

1. The reinforcement, corresponding to the ties considered in the design model, should be fully anchored beyond the nodes by bending up the bars, by using U-hoops or by anchorage devices, unless a sufficient length is available between the node and the end of the beam permitting an anchorage length of $l_{b,net}$.
2. Deep beams should normally be provided with a distributed reinforcement near both sides, the effect of each being equivalent to that of an orthogonal mesh with a reinforcement ratio of at least 0.15 per cent in both directions.

6.16   Anchorage Zones for Post-tensioned Forces (clause 5.4.6)

The approach to detailing the anchorage zone is different in each code. EC2 requires orthogonal mesh to be placed close to all surfaces. In addition, where groups of post-tensioned cables are located at a certain distance from each other, suitable links should be arranged at the ends of the members, as a protection against splitting.

At any part of the zone, the reinforcement ratio on either side of the block should be at least 0.15 per cent in both directions. All reinforcement should be fully anchored.

Where a strut and tie model has been used to determine the transverse tensile force, the following detailing rules should be followed:

1. The steel area actually required to provide the tie force, acting at its design strength, should be distributed in accordance with the actual tensile stress distribution, i.e. over a length of the block approximately equal to its greatest internal dimensions.
2. Closed stirrups should be used for anchorage purposes.
3. All the anchorage reinforcement should preferably be formed on to a three-dimensional orthogonal grid.

Special attention should be given to anchorage zones having cross-sections different in shape from that of the general cross-section of the beam.

BS 8110 requires two checks: one at serviceability limit state and the other at ultimate limit state (clauses 4.11.2, 4.11.3).

The bursting force, f_{bst}, is distributed in a region extending from $0.2y_0$ to $2y_0$ from the loaded face, and should be resisted by reinforcement in the form of spirals or closed links, uniformly distributed throughout this region and acting at a stress of 200 N/mm^2.

When a large block contains several anchorages it should be divided into a series of symmetrical loaded prisms and each prism treated in the above manner. However, additional reinforcement will be required around the groups of anchorages to ensure overall equilibrium of the end block. Special attention should also be paid to end blocks having a cross-section different in shape from that of the general cross-section of the beam.

For members with unbonded tendons the design bursting tensile force, F_{bst}, should be assessed from Table 4.7 of BS 8110 on the basis of the characteristic tendon force; the reinforcement provided to sustain this force may be assumed to be acting at its design strength ($0.87f_y$). No such check is necessary in the case of members with bonded tendons.

6.17   Reinforced Concrete Walls (clause 5.4.7)

The minimum and maximum area of vertical reinforcement is the same for EC2 and BS 8110 (0.4 and 4 per cent respectively). EC2 requires that half of this should be located in each face. The distance between adjacent vertical bars should not exceed twice the wall thickness or 300 m, whichever is less. These limitations do not exist in BS 8110.

EC2 requires that the horizontal bars be placed outside the vertical bars (between the vertical reinforcement and the nearest surface). It should not be less than 50 per cent of the vertical reinforcement. The spacing of horizontal bars should be $\not> 300$ mm and the diameter not less than one-quarter of that of the vertical bars. If the area of the load-carrying vertical reinforcement exceeds $0.02A_c$ this reinforcement should be enclosed by links according to clause 5.4.1.2.2.

BS 8110 rules for horizontal bars are similar where the vertical reinforcement is $< 0.02A_c$ horizontal reinforcement should be provided depending upon the characteristic strength of that reinforcement:

1. $f_y = 250 \, \text{N/mm}^2$: 0.30 per cent of concrete area;
2. $f_y = 460 \, \text{N/mm}^2$: 0.25 per cent of concrete area.

These horizontal bars should be evenly spaced and be not less than one-quarter of the size of the vertical bars and $\not< 6 \, \text{mm}$ (clause 3.1.2.7.4).

Where the vertical reinforcement exceeds $0.02A_c$, links at least 6 mm or one-quarter the size of the largest compression bar should be provided through the thickness of the wall. The spacing of links should not exceed twice the wall thickness in either the horizontal or vertical direction. In the vertical direction it should be $\not> 16$ times the bar size. Any vertical compression bar not enclosed by a link should be within 200 mm of a restrained bar.

6.18 Particular Cases (clause 5.4.8)
6.18.1 Concentrated Forces (clause 5.4.8.1)

EC2 requires the following special checks for which there are no corresponding clauses in BS 8110. Where one or more concentrated forces act at the end of a member or at the intersection of two structural members, local supplementary reinforcement should be provided capable of resisting the transverse tensile forces caused by these forces. This supplementary reinforcement may consist of links or of layers of reinforcement bent in the shape of hairpins.

For a uniform distribution of load on area A_{co} (Fig. 6.16), the concentrated resistance force can be determined as follows:

$$F_{Rdu} = A_{co}\alpha f_{cd}\sqrt{A_{c1}/A_{co}} \not> 3.3\alpha f_{cd}A_{co} \quad \text{(NAD)} \qquad [6.9]$$

where $f_{cd} = f_{ck}/\gamma_c$, A_{co} denotes the loaded area, A_{c1} denotes the maximum area corresponding geometrically to A_{co}, having the same centre of gravity, which it is possible to inscribe in the total area A_c, situated in the same plane as the loaded area, and α is a coefficient taking account of long-term effects on the compressive strength and unfavourable effects resulting from the way the load is applied ($= 0.85$). If A_c and A_{co} correspond geometrically and have the same

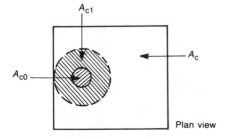

Plan view

Figure 6.16 Definition of the areas to be introduced in eq. 6.9

centre of gravity: $A_{c1} = A_c$. The value of F_{Rdu} obtained from Eq. 6.9 should be reduced if the load is not uniformly distributed on area A_{co} or if it is accompanied by large shear forces. This method does not apply to the anchorage of prestressing tendons.

6.18.2 Forces Associated with Change in Direction (clause 5.4.8.2)

At points where considerable changes in the direction of the internal forces occur, the associated radial forces shall be resisted by means of suitably anchored additional reinforcement or by detailing the normal reinforcement in a special way. BS 8110 does not have a corresponding clause.

6.18.3 Indirect Supports (clause 5.4.8.3)

EC2 requires that in the case of a connection between a supporting beam and a supported beam, 'suspension' reinforcement shall be provided and designed to resist the total mutual support reaction. The suspension reinforcement should consist preferably of links surrounding the principal reinforcement of the supporting member. Some of these links may be distributed outside the volume of concrete which is common to the two beams, as indicated in Fig. 5.20 of EC2.

BS 8110 requires that where the load on a beam is applied near the bottom of a section, sufficient vertical reinforcement to carry the load should be provided in addition to any reinforcement required to resist shear.

6.19 Limitation of Damage due to Accidental Actions (clause 5.5)

EC2 provides qualitative clauses compared with the specific requirements of BS 8110 (clause 3.12.3). The only specific requirement of EC2 is that continuity of the ties requires lap lengths equal to $2l_b$ and that the lap is enclosed by stirrups or spirals with $s \leq 100$ mm. In some cases continuity may be obtained by welding or by the use of mechanical connectors.

7 Prestressed Concrete

7.1 General

Prestressed concrete is treated differently in EC2 compared with BS 8110, the latter following the traditional British practice of treating prestressed concrete as a separate material. The Eurocode, on the other hand, takes the view that prestressed concrete is just part of a much wider material group known as reinforced concrete which covers normal reinforced concrete through partially prestressed concrete to fully prestressed concrete. The principles and methods given in EC2 are, in general, applicable to the full range of reinforced concrete construction. This chapter compares the two codes and highlights the differences between them as they affect the design of prestressed concrete members. Where such comparisons apply equally to normal reinforced concrete reference will be made to earlier chapters in this book rather than repeating the information.

EC2 Part 1 covers the design of prestressed concrete members using only bonded internal tendons. Part 1.5 of the code will contain special rules applicable to unbonded and external tendons. BS 8110 covers the design of members with either bonded or unbonded, internal or external tendons. At the time of writing, Part 1.5 of EC2 had not been published, therefore this chapter will consider only the design of members with internal bonded tendons.

7.2 Summary of Main Clauses

The main clauses relating to the design of prestressed concrete in EC2 and BS 8110 are summarized in Table 7.1.

7.3 Durability

Steel reinforcement and prestressing tendons are protected against corrosion by complying with code requirements on stress levels and crack widths. These are discussed in detail in section 7.5.2. In addition, both codes specify minimum values for the concrete cover to both reinforcement and prestressing tendons. The requirements in EC2 are for a 'minimum' cover to which a construction tolerance of 5 mm must be added, while BS 8110 is written in terms of 'nominal' covers which already include an allowance for the construction tolerance.

Table 7.1 Comparison of clauses for the design of prestressed concrete

Requirement	EC2	BS 8110
Durability		
Concrete cover	4.1.3.3.	4.12.3, 4.12.5.2
Design data		
Minimum concrete strength	4.2.3.5.2	4.1.8.1
Prestressing steel		
Mechanical properties	4.2.3.3	2.4.2.3
Minimum bending radii	4.2.3.3.6	4.7.2
Minimum number of wires	4.2.3.5.3	—
Jacking force	4.2.3.5.4	4.7.1
Ultimate limit state		
Bending and longitudinal force	4.3.1	4.3.7, 4.4, 4.5, 4.6
Values of prestress	2.3.3.1, 2.5.4.2, 2.5.4.4	—
Shear		
General	4.3.2	4.3.8, 4.4.1
Reduction in web width	4.3.2.2(8)	4.3.8.1
Shear capacity of concrete	4.3.2.3(1)	4.3.8.4, 4.3.8.5
Max. design shear stress	4.3.2.2(4), 4.3.2.3(3) 4.3.2.4.3(4), 4.3.2.4.4	4.3.8.2
Enhancement near supports	4.3.2.2(9), (10), (11)	—
Variable depth members	4.3.2.4.5	4.3.8.4, 4.3.8.5
Inclined tendons	4.3.2.4.6	4.3.8.4, 4.3.8.5
Torsion	4.3.3	4.3.9, Part 2
Serviceability limit state		
Values of prestress	2.5.4.2, 2.5.4.3	—
Stress levels	4.4.1	4.3.4.2, 4.3.5.1
Cracking	4.4.2	4.1.3, 4.3.4.3, 4.12.6
Shear reinforcement	4.4.2.3(5)	4.3.8.10
Deformation	4.4.3, Appendix 4	4.3.6.1
Torsion	4.4.2.3(5), 5.4.2.2, 5.4.2.3	—
Prestress losses		
Relaxation	4.2.3.4.1, 4.2.3.5.5(7), (9)	4.8.2
Elastic deformation	4.2.3.5.5(6)	4.8.3
Shrinkage	3.1.2.5.5, 4.2.3.5.5(9)	4.8.4
Creep	3.1.2.5.5, 4.2.3.5.5(9)	4.8.5
Draw-in	4.2.3.5.5(5)	4.8.6
Duct friction	4.2.3.5.5(8)	4.9
Anchorage zones		
Pretensioned members	4.2.3.5.6	4.10
Post-tensioned members	2.5.3.6.3, 2.5.3.7.4, 4.2.3.5.7, 5.4.6	4.11
Detailing		
Spacing of tendons/ducts	5.3.3	4.12.4, 4.12.5.3, 4.12.5.4
Anchorages and couplers	5.3.4	—
Minimum area of tendons	5.4.2.1.1	4.12.2
Tendon profile	5.4.2.1.3	—
Minimum shear reinforcement	5.4.2.2(5)	4.3.8.7
Spacing of shear reinforcement	5.4.2.2(7), (9)	4.3.8.10

For post-tensioned members, EC2 specifies that the minimum cover should not be less than the duct diameter. BS 8110 specifies a minimum nominal cover of 50 mm irrespective of duct diameter. Although EC2 specifies requirements for the minimum cover perpendicular to the plane of curvature for curved bars with yield strengths up to $500 \, N/mm^2$, it does not specify similar requirements for curved prestressing tendons. This is an unfortunate omission and the designer is referred to BS 8110 where specific requirements are stated for preventing bursting of the cover perpendicular to the plane of curvature.

For pretensioned members, EC2 requires the minimum cover to be twice the tendon diameter, or three times the diameter when ribbed wires are used, while BS 8110 specifies a nominal cover equal to the greater of the tendon diameter or the aggregate size.

In addition to these general requirements both codes specify minimum values for cover based on the condition of exposure to which the member will be subject. In EC2, the minimum covers are independent of concrete strength, while in BS 8110 they are related to concrete strength or, more correctly, to water/cement ratio and minimum cement content. However, there is a note in EC2 which instructs the designer to refer to Table 3 of ENV 206 in order to select the appropriate concrete quality to use with the specified covers. The cover and concrete quality requirements are compared in Table 7.2.

7.4   Design Data
7.4.1   Concrete

The minimum concrete strengths for use in pretensioned and post-tensioned construction are effectively the same in both codes. Differences between the design data to be assumed for concrete in each code affect prestressed members in the same way as they affect normal reinforced members and reference should be made to Chapter 4, where these are discussed.

7.4.2   Prestressing Steel

The stress−strain diagram for prestressing steel varies between the two codes, as shown in Fig. 7.1. EC2 adopts a bilinear diagram with either a horizontal or sloping top leg, while BS 8110 uses a trilinear diagram. The maximum design stress can be up to 10 per cent lower in EC2 so that when the ultimate limit state governs the design and the tendons are sufficiently highly stressed initially for their design strength to be reached at the ultimate limit state, EC2 will require approximately 10 per cent more tendons.

EC2 contains requirements on the minimum radii to which tendons of different composition can be bent. BS 8110 gives no guidance in this area, the designer having to rely on the tendon suppliers' literature. However, the British code does draw the designer's attention to the possible influence of the size of any deflector on the strength of the tendon.

7.2 Comparison of concrete quality and cover to prestressing tendons or ducts (mm)

Exposure class		Cover			Max. water/ cement ratio	Min. cement content (kg/m³)	Lowest grade of concrete
	BS 8110	Min.	Nominal†	NAD‡			
1	—	25	30	25	0.60	300	—
	Mild		20		0.60	300	28/35
2a	—	30	35	40	0.60	300	—
	Moderate		20		0.45	400	40/50
	Moderate		25		0.50	350	35/45
	Moderate		30		0.55	325	32/40
	Moderate		35		0.60	300	28/35
2b	—	35	40	40	0.55	300§	—
	Severe		25		0.45	400	40/50
	Severe		30		0.50	350	35/45
	Severe		40		0.55	325	32/40
3		50	55	45	0.50	300§	—
4a		50	55	45	0.55	300	—
4b		50	55	45	0.50	300§	—
5a		35	40	40	0.55	300	—
5b		40	45	40	0.50	300	—
5c		50	55	50	0.45	300	—
	Very severe		30		0.45	400	40/50
	Very severe		40		0.50	350§	35/45
	Very severe		50		0.55	325§	32/40
	Extreme		50		0.45	400	40/50
	Extreme		60		0.50	350§	35/45

† Includes a construction tolerance of 5 mm.
‡ Nominal covers as specified in the UK NAD.
§ Air entrainment required.

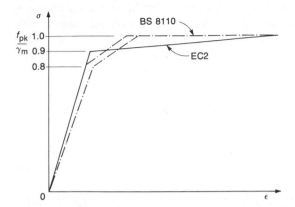

Figure 7.1 Comparison of stress–strain curves for prestressing tendons

EC2 addresses the important issue of reliability by including simple requirements on the minimum number of bars, wires and tendons in the precompressed zone of isolated members, i.e. 'Members in which no additional load-carrying capacity due to redistribution of internal forces and moments, transverse redistribution of loads or due to other measures (e.g. normal steel reinforcement) exists'. Seven-wire strands meet the reliability requirement, provided that the minimum wire diameter is greater than 4 mm. Therefore, there should be no effect on normal British practice.

The maximum initial stress applied to the tendon, i.e. the jacking stress, is expressed as a percentage of the ultimate strength of the tendon in BS 8110 and of either the ultimate strength (f_{pk}) or the 0.1 per cent proof stress $(f_{p0.1k})$ in EC2. In practice, $f_{p0.1k} \approx 0.85 f_{pk}$ and the requirement expressed in terms of the 0.1 per cent proof stress will be the more critical. EC2 permits the use of a higher jacking load than BS 8110 when $f_{p0.1k} > 0.83 f_{pk}$. Similar requirements apply to the stress in the tendon after transfer. Again, EC2 permits the use of higher loads when $f_{p0.1k} > 0.83 f_{pk}$, or $0.88 f_{pk}$ if advantage is taken of the maximum allowable value in BS 8110. Table 7.3 shows that EC2 permits approximately

Table 7.3 Comparison of jacking loads for different strand types (kN)

Strand	Dia. (mm)	f_{pk} (N/mm²)	$f_{p0.1k}$ (N/mm²)	EC2 Jacking stress	EC2 Initial prestress	BS 8110 Jacking stress	BS 8110 Initial prestress
STD	15.2	1670	1420	1278	1207	1253 (1336)	1169 (1253)
	≤ 12.5	1770	1500	1350	1275	1328 (1416)	1239 (1328)
SUP	15.7	1770	1500	1350	1275	1328 (1416)	1239 (1328)
	≤ 12.9	1860	1580	1422	1343	1395 (1488)	1302 (1395)
DYF	18.0	1700	1450	1305	1233	1275 (1360)	1190 (1275)
	15.2	1820	1545	1391	1313	1365 (1456)	1274 (1365)
	12.7	1860	1580	1422	1343	1395 (1488)	1302 (1393)

Note: Values in parentheses ( ) are based on the higher values allowed in BS 8110.

3 per cent higher jacking loads. In practice, the effect of such a difference is difficult to determine because the required area of prestressing tendons depends not only on the initial prestress but also on the variations between the two codes in the level of predicted prestress losses and in the design criteria at the ultimate and serviceability limit states. This is discussed further in section 7.7.

7.5   Design of Sections for Flexure and Axial Load

7.5.1   Ultimate Limit State

The effect of differences between the codes in the load factors and in the stress−strain relationship for concrete will be similar to that discussed in Chapter 4 for normal reinforced concrete.

Both codes adopt a partial safety factor for prestressing force (γ_p) of 1.0 for the analysis of structures. For the design of sections, BS 8110 implicitly assumes that $\gamma_p = 1.0$ in all cases, while EC2 takes $\gamma_p = 1.0$ provided that the following conditions are both met:

1. Not more than 25 per cent of the total area of prestressed steel is located within the compression zone at the ultimate limit state;
2. The stress at the ultimate limit state in the prestressing steel closest to the tension face exceeds $f_{p0.1k}/\gamma_m$.

If these conditions are not met, EC2 adopts a value of $\gamma_p = 0.9$, which is to take account of variations across the section of the prestrain in the prestressing steel corresponding to a concrete stress $\sigma_c = 0$. These could be significant for sections near to points of contraflexure where, because these sections are subjected to very low bending moments, the tendons may be spread across the section and experience a considerable variation in stress due to applied loads.

In practice, most prestressed members will meet both conditions (1) and (2) so that the only difference between the two codes will be due to the lower maximum design tendon stress allowed in EC2, see section 7.4.2.

7.5.2   Serviceability Limit State

Both codes cover the three most common serviceability limit states, i.e. maximum concrete compressive stress, crack control and deformation.

EC2 distinguishes between three serviceabililty load combinations, namely rare, frequent and quasi-permanent. All are relevant in the design of prestressed concrete members. BS 8110 uses only one serviceability combination which corresponds to the rare combination in EC2. The relative magnitude of the three EC2 combinations depends on the loads applied and the type of structure being designed.

Comparison of the two codes is further complicated by the fact that BS 8110 always uses a mean characteristic value of the prestress while EC2 uses an upper or lower characteristic value (whichever is more critical) when checking crack widths and tendon stresses. The upper and lower characteristic values are taken as 1.1 and 0.9 times the mean value of the prestress respectively, as long as the sum of the friction and long-term losses is < 30 per cent. If this condition is not met, EC2 does not offer any suggestions as to what values the designer should adopt. It is suggested that the designer refers to section 4.2.6.2 of the draft CEB−FIP Model Code 1990.

Table 7.4 compares the different serviceability criteria in the two codes for a typical relationship between the different load combinations.

7.5.2.1 Limitation on Stress

To avoid longitudinal cracks occurring in regions of high compression, EC2 limits the compressive stress under the rare combination to $0.6f_{ck}$ in areas exposed to environments of exposure class 3 or 4 unless other measures, such as an increase in cover to reinforcement in the compressive zone or confinement by transverse reinforcement, are taken. In order to ensure that creep deformation is within the limits predicted by other parts of EC2, the concrete stress under the quasi-permanent combination is limited to $0.45f_{ck}$. BS 8110 covers both situations by limiting compressive stresses to $0.25f_{cu}$ for uniform stress distributions and $0.33f_{cu}$ or $0.40f_{cu}$ for triangular distributions. Taking account of the different load combinations, EC2 is generally less stringent than BS 8110, see Table 7.4.

Table 7.4 Comparison of allowable stresses for the serviceability limit state (N/mm^2)

Concrete grade	Code	Combination		
		Rare	Frequent	Quasi-permanent
Combination factor		1.0	0.85	0.75
30/37	EC2 compression	18.0 $(\gamma_p=1)$	—	18.0 $(\gamma_p=1)$
	BS 8110 compression	12.2	—	—
	Class 1	0	—	—
	Class 2	−2.2	—	—
	Class 3	−4.6	—	—
40/50	EC2 compression	24.0 $(\gamma_p=1)$	—	24.0 $(\gamma_p=1)$
	BS 8110 compression	16.5	—	—
	Class 1	0	—	—
	Class 2	−2.6	—	—
	Class 3	−5.8	—	—
50/60	EC2 compression	30.0 $(\gamma_p=1)$	—	30.0 $(\gamma_p=1)$
	BS 8110 compression	19.8	—	—
	Class 1	0	—	—
	Class 2	−2.8	—	—
	Class 3	−5.8	—	—
Tendons	EC2 tension $(\gamma_p=0.9$ or $1.1)$	−0.75f_{yk}	Cracking/ decompression	

Notes: (1) The combination factor expresses the combination as a fraction of the rare combination. (2) To allow the criteria to be compared, allowable stresses are calculated by dividing the actual allowable stress by the combination factor.

BS 8110 allows higher fractions of the cube strength to be used at transfer (up to $0.5f_{ci}$, where f_{ci} is the concrete strength at the time of transfer). EC2 suggests that it may be necessary to limit the concrete compressive stress at transfer to $0.45f_{ck}$ (f_{ck} being based on the concrete strength at transfer), if creep at this stage is likely to affect significantly the functioning of the member. Otherwise, it is suggested that compressive stresses at transfer are limited to $0.6f_{ck}$.

To avoid stresses in the tendons under serviceability conditions which could lead to inelastic deformation of the tendons, EC2 limits the stress to $0.75f_{pk}$, after allowance for losses. This is unlikely to be a critical criterion is most cases and there is no equivalent requirement in BS 8110.

EC2 draws the designer's attention to the fact that in partially prestressed members creep and shrinkage can lead to high stresses in both normal reinforcement and prestressing tendons which could result in fatigue problems. This is a useful reminder to the designer.

7.5.2.2 Cracking

BS 8110 defines three classes of prestressed concrete member which depend on the level of flexural tensile stress allowed. These are as follows:

Class 1. No flexural tensile stresses;
Class 2. Flexural tensile stresses but no visible cracking;
Class 3. Flexural tensile stresses with crack widths limited to specific values, generally 0.1 mm or 0.2 mm.

EC2 effectively specifies only two classes — decompression and limiting crack widths to 0.2 mm. There are two important differences between the two codes. In EC2, the decompression limit requires all parts of the tendon or duct to lie at least 25 mm within concrete in compression compared with the class 1 requirement of BS 8110 for the whole concrete section to remain in compression. All cracking checks in EC2 for prestressed concrete are carried out under the frequent load combination using the upper or lower characteristic prestressing force rather than the equivalent of the rare combination and the mean prestressing force specified by BS 8110.

In order to limit crack widths, EC2 tabulates bar size and spacing criteria to ensure that widths are limited to 0.2 mm, as long as a minimum area of reinforcement or prestressing tendons is provided to control cracking due to the restraint of imposed deformations. Alternatively, formulae are provided to allow the design crack width to be calculated. The minimum area of reinforcement to control cracking arising from the restraint of imposed deformations can be reduced by taking account of the contribution of the prestressing tendons. BS 8110 adopts a different approach, specifying allowable hypothetical tensile stresses in the concrete which have been selected to meet the requirements of either class 2 or class 3. This approach was first adopted in CP 110 when it was introduced in 1972 because there was no generally accepted crack width formula for prestressed concrete at that time.

The approach adopted by EC2 for the design of partially prestressed concrete may be more rigorous, but it is more complex and difficult for the designer to follow. Firstly, it should be noted that Tables 4.11 and 4.12 of the code apply only to high bond bars. Prestressing strands have bond characteristics which are significantly less effective than high bond bars, according to EC2, clause 4.4.2.4(4). It is theoretically possible to use Table 4.11 by determining an equivalent high bond bar diameter using the following formula:

$$\phi' = \frac{k_{1ps}}{k_{1hb}} \times \phi$$

where ϕ' is the equivalent high bond bar diameter, ϕ the prestressing strand diameter, k_{1hb} the value for k_1 for high bond bars (0.8), given in clause 4.4.2.4(3) and k_{1ps} the value of k_1 for prestressing strand (2.0), given in clause 4.4.2.4(4). However, this will give equivalent diameters outside the range of Table 4.11 for most normal strand diameters. The designer is therefore forced to use the crack width formulae. The procedure to be followed is as follows:

1. Determine the moments for the frequent combination and the design value of the prestress force ($0.9P_{m,\infty}$).
2. Calculate the cracking moment and hence determine σ_{sr}, the stress in the tension reinforcement calculated on the basis of a cracked section. The code does not explicitly state whether the lower (5 per cent) fractile value or the mean value of the flexural tensile strength of concrete should be used for this purpose. As the aim is to limit the size of any cracks, it would seem reasonable to adopt the lower value of flexural tensile strength. Once it is accepted that the section is cracked, the difference between the two values is unlikely to be of any significance as it only affects the degree of tension stiffening taken into account.
3. Calculate the neutral axis depth, x, and the stress in the tension reinforcement under the frequent load combination, σ_s, for the cracked section.
4. Calculate the mean strain, ϵ_{sm}, using Eq. 4.81 of the code. Here β_1 should be taken as 0.5 for prestressing strand and β_2 as 0.5 for frequent loading.
5. Calculate the average final crack spacing, s_{rm}, from Eq. 4.82. Here k_1 should be taken as 2.0 for prestressing strand and k_2 as 0.5 for bending. Note that s_{rm} is limited to $(h - x)$ in clause 4.4.2.4(8), where h is the overall depth of the section.
6. Calculate the design crack width, w_k, using Eq. 4.80.
7. Compare w_k with the required value and adjust the prestressing force and/or its eccentricity.
8. Repeat steps 1–7 until the required value of w_k is obtained.

In prestressed concrete sections which also contain ordinary reinforcement, the latter can be used to control crack widths rather than relying on the prestress. In this case, Tables 4.11 and 4.12 of EC2 will be applicable.

It is difficult to draw general conclusions regarding the effect on the final prestress required by the different approaches of the two codes because of the number of variable parameters. However, the examples shown in Figs 7.2 and 7.3 serve to illustrate some of these effects.

The areas of prestressing tendons required for different cracking criteria are summarized in Table 7.5. The EC2 decompression criterion requires only 95 per cent of the tendon area needed to meet the class 1 requirements of BS 8110. When crack widths are limited to 0.2 mm, EC2 requires between 15 and 50 per cent more area than BS 8110, depending on the shape of the cross-section. However, the area required by BS 8110 must be increased to provide the necessary capacity at the ultimate limit state. For the T-section the area required by EC2 approaches that for a class 2 design to BS 8110.

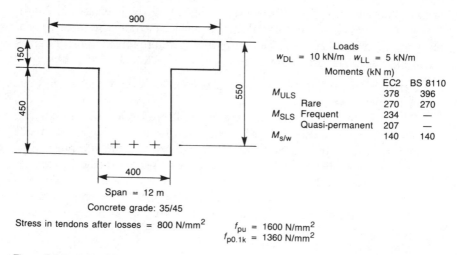

Figure 7.2 Prestressed T-beam example

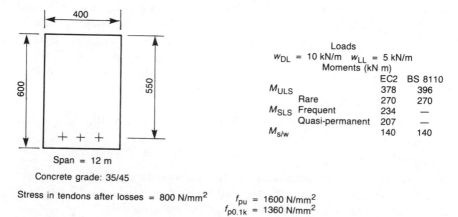

Figure 7.3 Prestressed rectangular beam example

Table 7.5 Areas of prestressing tendons for various limit states (mm^2)

Code	ULS	SLS		
		Class 1/ decompression	Class 2	Class 3 (0.2 mm)
T-beam†				
BS 8110	542	856	635	411
EC2	525	815	—	599
Rectangular beam†				
BS 8110	569	965	758	546
EC2	550	915	—	638

† For cross-sections see Figs 7.2 and 7.3.

7.5.2.3 Deformation

BS 8110 specifically warns the designer that the use of span/depth ratios is not appropriate for limiting deflections of prestressed members because of the major influence of the level of prestress. If the determination of deflection is necessary, the designer is advised to use elastic methods adopting design criteria and material properties from Part 2 of the code. EC2 does not contain the same specific warning, but the span/depth ratios given in Table 4.14 of the code are stated to apply only to reinforced concrete. Guidance is given in Appendix 4 of the code on the calculation of deformations using elastic methods.

7.5.3   Effect on the Design of Prestressed Members

Because of the interrelationship between the different parameters which determine the design of a prestressed concrete member, it is not easy to compare the effects of individual parameters. Therefore, two design examples have been carried out for the sections shown in Figs 7.2 and 7.3, assuming that the losses in both the EC2 designs and the BS 8110 designs are the same. The results are summarized in Table 7.5 and typical Magnel diagrams are presented in Figs 7.4 and 7.5 for designs to classes 1 and 3.

The Magnel diagrams show that, for class 1 members, EC2 requires a smaller tendon area. BS 8110 permits the use of a higher tendon force at transfer (if the concrete compressive stress at transfer is limited to $0.45f_{ck}$ in EC2), except at high eccentricities when tensile stresses will be induced in the top fibres. For class 3 members, designs to EC2 are governed by the ultimate limit state for small eccentricities and by the crack width criterion at higher eccentricities. Class 3 designs to BS 8110 tend to be governed by the ultimate limit state for all eccentricities. BS 8110 will lead to more economical designs at high eccentricities, but there is little difference between the two codes at small eccentricities. Class 2 members designed to BS 8110 lie between class 1 and class 3 members and any comparison with EC2 designs will depend on whether the latter adopts the

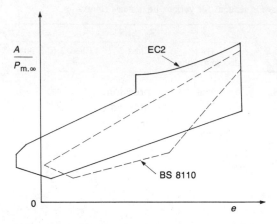

Figure 7.4 Comparison of tendon zones, T-beam, class 1

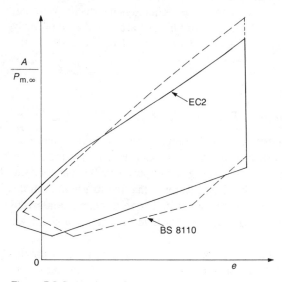

Figure 7.5 Comparison of tendon zones, T-beam, class 3 (crack width = 0.2 mm)

decompression or crack width criterion. Class 2 members to BS 8110 are more economical at high eccentricities than members designed to the EC2 decompression limit, but this situation is reversed if a limiting crack width of 0.2 mm is adopted. At low eccentricities, which would occur close to points of contraflexure in continuous members, EC2 always requires a smaller tendon area. The Magnel diagrams and the conclusions which can be drawn from them are similar for both the sections analysed.

7.6   Design of Sections for Shear and Torsion

7.6.1   Shear

7.6.1.1 Ultimate Limit State

There is a fundamental difference in the way the two codes approach the design of sections for shear. EC2 follows the same procedure as for reinforced concrete, treating the prestressing force as an applied axial compression. BS 8110 adopts a different approach from that which it follows for reinforced concrete. Checks are carried out assuming that the section is first uncracked and then (if appropriate) cracked, and the lower value is taken as the shear capacity of the section.

Although there is this difference in approach, there are a number of aspects in which the two codes are similar. Minimum shear reinforcement is not required by either code in slabs having adequate provision for the transverse distribution of loads or in members of minor importance, e.g. a lintel with a span of less than 2 m. Both codes require a reduction in web width to be taken into account when webs contain ducts. The detailed requirements in the two codes are slightly different, BS 8110 requiring the greater reduction. In EC2, this reduction is only required when checking the maximum shear force which can be carried without crushing the concrete.

BS 8110 determines the shear capacity of an uncracked concrete section by limiting the principal tensile stress at the section, using a partial load factor of 0.8 on the prestress force. For a cracked section, the shear capacity is calculated using an empirical equation:

$$V_{cr} = \left(1 - 0.55 \frac{f_{pe}}{f_{pu}}\right) v_c b_v d + M_0 \frac{V}{M}$$

The second term of the equation is effectively the shear force carried by the section when the extreme tensile fibre becomes zero, i.e. the bending moment (M) reaches a value M_0, while the first term is the additional shear force which can be carried by the cracked concrete section before failure occurs. Once again, a partial safety factor of 0.8 is applied to the prestressing force in calculating the moment, M_0. The total shear capacity of the section is the sum of the shear capacity of the concrete section alone and the shear capacity of the reinforcement.

In EC2, the shear capacity of a prestressed concrete member without shear reinforcement is determined using the same equation as for reinforced concrete, with the prestressing force being treated as an axial load. A partial safety factor, $\gamma_p = 1.0$, should be used unless the conditions given in clause 2.5.4.4.3(3) are not satisfied, when γ_p should be reduced to 0.9. If the area of prestressing tendons has been increased to satisfy the serviceability limit state for flexure or the ultimate moment coincident with the design shear force is significantly less than the ultimate design moment, the stress at the ultimate limit state in the tendons closest to the tensile fibre may not exceed $f_{p0.1k}/\gamma_m$ and γ_p must be taken as 0.9.

One of the factors governing the shear capacity of the concrete is the area of longitudinal reinforcement in the tensile zone. BS 8110 clearly states that this should be the sum of both the normal reinforcement and the prestressing tendons, with no adjustment for the different design strengths. EC2 is silent on this point, but the approach in BS 8110 should be adopted as this term is included to account for the change in level of the neutral axis which is governed more by reinforcement area rather than by its strength.

The shear resistance of concrete according to EC2 and the shear capacity of an uncracked section to BS 8110 are compared in Table 7.6. This shows that the shear resistance to EC2 is significantly less than that to BS 8110, varying from just under 50 per cent of the BS 8110 value for low levels of prestress to 90 per cent at high levels of prestress. Therefore, in sections which are uncracked at the ultimate limit state, the shear reinforcement required by BS 8110 will be much less than that required by EC2. Table 7.6 also shows the minimum shear capacity of a cracked section to BS 8110.

Table 7.6 Comparison between allowable shear stresses for EC2 and an uncracked section to BS 8110 (N/mm^2)

Mean prestress (N/mm^2)	Concrete grade							
	25/30		30/37		40/50		50/60	
	EC2	BS 8110	EC2	BS 8110	EC2	BS 8110	EC2	BS 8110
Min. value for cracked section		0.55		0.61		0.71		0.77
2	0.60	1.30	0.65	1.40	0.75	1.60	0.75	1.70
4	0.90	1.65	1.00	1.75	1.05	1.95	1.05	2.05
6	1.20	1.90	1.30	2.05	1.40	2.20	1.40	2.35
8	1.50	2.15	1.60	2.25	1.70	2.50	1.70	2.65
10	1.80	2.35	1.90	2.50	2.00	2.70	2.00	2.85
12	2.10	2.55	2.20	2.70	2.30	2.95	2.30	3.10
14	2.40	2.70	2.50	2.90	2.60	3.15	2.60	3.30

EC2 allows the designer the choice of two methods for determining the shear capacity of a section with shear reinforcement. Both adopt a truss model for determining the capacity of the shear reinforcement. In the standard method, the concrete compression struts are assumed to act at 45° to the horizontal and the shear capacity of the concrete section is added to the capacity of the shear reinforcement to determine the total capacity of the section. This is similar to the approach adopted by BS 8110. The formulae for the capacity of the reinforcement in EC2 and BS 8110 are similar, the latter requiring about 10 per cent less reinforcement as it is based on the effective depth, d, rather than the lever arm which is taken to be $0.9d$ in EC2.

In the variable strut inclination method, the designer is free, within certain limits, to choose the angle of the concrete compression strut. The area of shear reinforcement required to maintain equilibrium can then be calculated. In this method the shear capacity of the concrete is not added to the shear capacity of the truss, and consequentially no benefit is gained from prestressing in the horizontal direction.

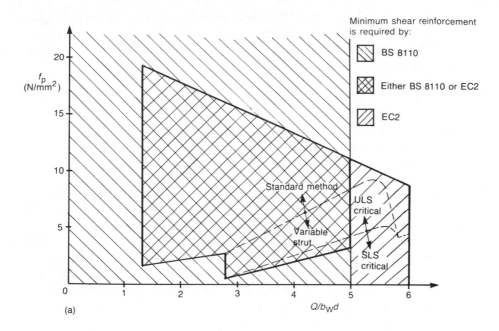

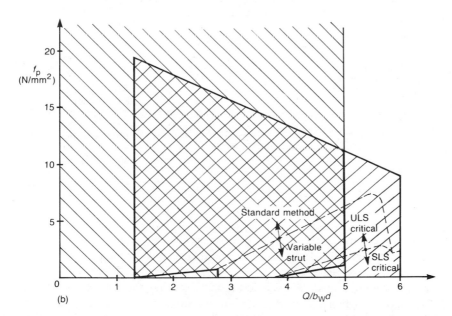

Figure 7.6 Comparison of shear reinforcement, concrete grade 40/50. (a) Longitudinal reinforcement ratio, $\rho = 0.0015$; (b) longitudinal reinforcement ratio, $\rho = 0.02$

However, this method can still lead to a more economical design of shear reinforcement in some situations as demonstrated in Fig. 7.6.

EC2 specifically draws the designer's attention to the fact that the tensile force in the longitudinal reinforcement is increased by the presence of shear above that required for bending. Provision for this effect is generally made by shifting the bending moment diagram·so that the moment at a given section is always increased. This is discussed in the section on detailing. It should be noted that, although the variable strut inclination method can lead to lower areas of shear reinforcement, there will be a corresponding increase in the longitudinal reinforcement requirement.

Both codes limit the maximum shear force which can be carried by the section to that which will cause crushing of the concrete. In BS 8110, this limit is independent of the prestressing force, while in EC2, the capacity is reduced when the section is subjected to an axial compressive force. Although not specifically stated in the code, it is suggested that this reduction should also apply to any section subjected to a prestressing force. In this situation the prestressing force is an unfavourable effect and a partial safety factor of $\gamma_p = 1.2$ should be used. Table 7.7 compares the maximum allowable shear stresses in each code and shows that without such a reduction the maximum shear capacity would be significantly greater in EC2 which, unlike BS 8110, does not have an upper limit to the maximum shear capacity.

Table 7.7 Comparison of maximum allowable shear stresses (N/mm²)

Code	Mean prestress (N/mm²)	Concrete grade					
		25/30	30/37	35/45	40/50	45/55	50/60
EC2							
$V_{Rd2.red}/b_w d$	0	4.31	4.95	5.51	6.00	6.75	7.50
	2	4.31	4.95	5.51	6.00	6.75	7.50
	4	4.31	4.95	5.51	6.00	6.75	7.50
	6	4.09	4.95	5.51	6.00	6.75	7.50
	8	3.05	4.30	5.42	6.00	6.75	7.50
	10	2.02	3.31	4.47	5.51	6.75	7.50
	12	—	2.31	3.52	4.61	5.86	7.11
	14	—	—	2.58	3.71	4.96	6.21
BS 8110							
$V_{max}/b_v d$		4.38	4.87	5.00	5.00	5.00	5.00

A direct comparison between the shear resistance according to EC2 and that of a cracked section to BS 8110 is not possible because the approach adopted by the British code depends on the magnitude of the shear force and moment acting on the section. In order to demonstrate the effect of the differences between the codes, the variation of shear resistance of the beams considered in the examples of Figs 7.2 and 7.3 are shown in Tables 7.8 and 7.9 respectively. The design shear force and corresponding moment are also given. These results confirm those

Table 7.8 Comparison of concrete shear resistance for a T-beam

Distance from support (m)	0	1	2	3	4	5	6
BS 8110							
V (kN)	132	110	89	69	49	30	12
M (kN m)	0	117	207	270	309	327	324
Class 1							
M_0 (kN m)	216	216	216	216	216	216	216
V_{c0} (kN)	373	373	373	373	373	373	373
V_{cr} (kN)		289	179	148	148	148	148
V_c (kN)	373	373	373	148	148	148	148
A_{sv}/s_v (mm^2/m)	0	0	0	0	0	0	0
Class 2							
M_0 (kN m)	161	161	161	161	161	161	161
V_{c0} (kN)	347	347	347	347	347	347	347
V_{cr} (kN)		229	148	148	148	148	148
V_c (kN)	347	347	148	148	148	148	148
A_{sv}/s_v (mm^2/m)	0	0	0	0	0	0	0
Class 3							
M_0 (kN m)	104	104	104	104	104	104	104
V_{c0} (kN)	319	319	319	319	319	319	319
V_{cr} (kN)		165	148	148	148	148	148
V_c (kN)	319	165	148	148	148	148	148
A_{sv}/s_v (mm^2/m)	0	400	400	0	0	0	0
EC2							
V_{Sd} (kN)	126	105	85	66	47	29	11
M_{Sd} (kN m)	0	114	200	261	299	316	313
Class 1							
V_{Rd1} (kN)	177	177	177	177	177	177	177
A_{sw}/s (mm^2/m)	480	480	480	480	480	480	480
Class 3							
V_{Rd1} (kN)	157	157	157	157	157	157	157
A_{sw}/s (mm^2/m)	480	480	480	480	480	480	480

Note: For cross-section see Fig. 7.2.

shown in Table 7.6 for sections uncracked in flexure and show that the BS 8110 capacity is approximately twice that calculated from EC2. The capacity of the cracked concrete section calculated according to BS 8110 is lower than that for the uncracked section. As the moment on the section increases, this capacity quickly becomes less than the EC2 value and reaches its minimum value of $0.1b_v d\sqrt{f_{cu}}$, which is between 70 and 95 per cent of the EC2 capacity. In both Tables 7.8 and 7.9, a partial safety factor on the prestressing force (γ_p) of 0.9 has been used as the ultimate moments for the load cases giving the maximum design shear forces are such that the stress in the tendon nearest to the tensile fibre does not reach a value of $f_{p0.1k}/\gamma_m$.

A further comparison between the two codes is presented numerically in Table 7.10 and shown graphically in Fig. 7.6. The range of shear reinforcement ratios shown for the 'variable strut' covers the range of strut inclination angles specified in the UK NAD. The two values shown for BS 8110 correspond to the shear

Table 7.9 Comparison of concrete shear resistance for a rectangular beam

Distance from support (m)	0	1	2	3	4	5	6
BS 8110							
V (kN)	132	110	89	69	49	30	12
M (kN m)	0	117	207	270	309	327	324
Class 1							
M_0 (kN m)	216	216	216	216	216	216	216
V_{c0} (kN)	417	417	417	417	417	417	417
V_{cr} (kN)		292	182	148	148	148	148
V_c (kN)	417	417	417	148	148	148	148
A_{sv}/s_v (mm^2/m)	0	0	0	0	0	0	0
Class 2							
M_0 (kN m)	170	170	170	170	170	170	170
V_{c0} (kN)	389	389	389	389	389	389	389
V_{cr} (kN)		241	155	148	148	148	148
V_c (kN)	389	389	155	148	148	148	148
A_{sv}/s_v (mm^2/m)	0	0	400	0	0	0	0
Class 3							
M_0 (kN m)	122	122	122	122	122	122	122
V_{c0} (kN)	357	357	357	357	357	357	357
V_{cr} (kN)		190	148	148	148	148	148
V_c (kN)	357	357	148	148	148	148	148
A_{sv}/s_v (mm^2/m)	0	0	400	0	0	0	0
EC2							
V_{Sd} (kN)	126	105	85	66	47	29	11
M_{Sd} (kN m)	0	114	200	261	299	316	313
Class 1							
V_{Rd1} (kN)	207	207	207	207	207	207	207
A_{sw}/s (mm^2/m)	480	480	480	480	480	480	480
Class 3							
V_{Rd1} (kN)	176	176	176	176	176	176	176
A_{sw}/s (mm^2/m)	480	480	480	480	480	480	480

Note: For cross-section see Fig. 7.3.

reinforcement required for an uncracked section and a section with the minimum cracked shear capacity. Over the majority of the practical range it is not possible to draw a general conclusion as to which code requires the smaller area of shear reinforcement.

Although BS 8110 allows for the increased resistance of sections close to a direct support due to the direct transmission of loads for reinforced concrete sections, there are no specifically stated similar rules for prestressed concrete. It must therefore be concluded that such enhancements are not allowed. On the other hand, EC2 allows for an increased resistance in prestressed concrete in the same way as for reinforced concrete. However, no such enhancement should be considered when checking the design shear force against the maximum shear capacity of the section (V_{Rd2} or $V_{Rd2.red}$).

The effect on the design shear force (positive or negative) of inclined tendons and compression and tension zones must be included when calculating the design

Table 7.10 Comparison of shear capacity — concrete grade 40/50

$Q/b_w d$ (N/mm^2)	f_p (N/mm^2)	$A_{sw}/b_w s$		
		Variable strut	Standard	BS 8110
0	0	0.0015	0.0015	0.0000
	5	0.0015	0.0015	0.0000
	10	0.0015	0.0015	0.0000
	15	0.0015	0.0015	0.0000
	20	0.0015	0.0015	0.0000
1	0	0.0019 − 0.0041	0.0015	0.0010
	5	0.0019 − 0.0041	0.0015	0.0000−0.0010
	10	0.0019 − 0.0041	0.0015	0.0000−0.0010
	15	0.0019 − 0.0041	0.0015	0.0000−0.0010
	20	0.0028	0.0015	0.0000−0.0010
2	0	0.0037 − 0.0083	0.0033 − 0.0041	0.0018−0.0032
	5	0.0037 − 0.0083	0.0015 − 0.0022	0.0010−0.0032
	10	0.0037 − 0.0083	0.0015	0.0010−0.0032
	15	0.0037 − 0.0083	0.0015	0.0010−0.0032
	20	†	†	0.0010−0.0032
3	0	0.0056 − 0.0124	0.0061 −[0.0075]	0.0043−0.0057
	5	0.0056 − 0.0124	0.0042 − 0.0050	0.0017−0.0057
	10	0.0056 − 0.0124	0.0023 − 0.0031	0.0010−0.0057
	15	0.0056 − 0.0124	0.0015	0.0010−0.0057
	20	†	†	0.0010−0.0057
4	0	[0.0081]− 0.0166	0.0088 −[0.0128]	0.0068−0.0082
	5	0.0074 − 0.0166	0.0070 − 0.0078	0.0042−0.0082
	10	0.0074 − 0.0166	0.0051 − 0.0059	0.0024−0.0082
	15	†	†	0.0010−0.0082
	20	†	†	0.0010−0.0082
5	0	[0.0133]−[0.0181]	[0.0133]−[0.0181]	0.0093−0.0107
	5	0.0093 − 0.0167	0.0097 − 0.0106	0.0067−0.0107
	10	0.0093 − 0.0167	0.0079 − 0.0087	0.0049−0.0107
	15	†	†	0.0035−0.0107
	20	†	†	0.0023−0.0107
6	0	[0.0186]−[0.0234]	[0.0186]−[0.0234]	†
	5	0.0167	0.0125 − 0.0133	†
	10	†	†	†
	15	†	†	†
	20	†	†	†

† Shear capacity exceeded.
Note: Areas governed by crack control criteria are shown in parentheses [ ].

shear force in EC2 and also in BS 8110 at a section uncracked in flexure. However, when considering a cracked section under BS 8110, the design shear force should only be modified when the effect of inclined tendons or compression and tensile zones increases the design shear force. BS 8110 is conservative in this respect, basing its conservatism on a few limited test results, which appear to deny statics. When using EC2, a reduction in the design shear force due to inclined compression or tension zones can only be combined with a reduction due to inclined prestressing tendons if a detailed verification can be given. The appropriate partial safety factor to be used with EC2 depends on the stress in the tendons and on whether the effect of the load is favourable or unfavourable. For prestressing tendons the appropriate values are as follows:

1. When the stress in the tendons does not exceed $f_{p0.1k}$, $\gamma_p = 0.9$ for favourable effects and 1.2 for unfavourable effects;
2. When the stress in the tendons exceeds $f_{p0.1k}$ the code specifies that the prestressing force (V_{pd}) should be calculated assuming a stress of $f_{p0.1k}/\gamma_s$, which presumably applies for favourable effects. It does not specify the value to be used when the effect is unfavourable, and it is suggested that a value of $\gamma_p = 1.2$ is applied to the calculated force.

7.6.1.2 Serviceability

Although both codes design for shear at the ultimate limit state, only EC2 has specific requirements to control cracking due to shear forces. The code explains that cracking under service loads will only occur when the ultimate design shear force is greater than three times the shear capacity of the concrete section without shear reinforcement. It relates the excess stress in the shear reinforcement at the ultimate limit state to the maximum stirrup spacing. This criterion is likely to be critical in members with a high applied ultimate shear stress and a low mean value of prestress, as illustrated in Fig. 7.6 and Table 7.10.

7.6.2   Torsion

Both codes require calculations of torsional resistance to be carried out when the equilibrium of the structure depends on its torsional resistance. EC2 requires the design to be carried out at both the serviceability and ultimate limit states and provides the necessary design rules. The reinforcement is designed using the variable strut inclination method and this method must also be used when designing for any coexistent shear forces. BS 8110 requires torsion to be considered at the ultimate limit state only. The design rules are included in Part 2 of BS 8110, rather than in the main code.

When torsion arises from consideration of compatibility only, EC2 requires the member to be designed to avoid excessive cracking. In practice this means that stirrups and longitudinal reinforcement should satisfy specified detailing rules. Although not specifically stated in BS 8110, it is assumed that proper detailing will limit the width of any cracks which form.

7.7   Prestress Losses
7.7.1   General

Both codes draw the designer's attention to the need to allow for losses of prestress when calculating the design forces in tendons at the various stages considered in the design. The causes of these losses are listed. Because of the uncertainty in estimating such losses, both codes suggest that experimental evidence should be used where it is available. In the absence of such data, both codes suggest values for the various parameters which can be used for design.

7.7.2   Friction in Jack and Anchorages

Loss of prestressing force due to friction in the jack and anchorages is mentioned in BS 8110, but not in EC2. It is not generally a problem for the designer because his design is normally based on the tendon force on the concrete member side of the jack. However, it does need to be allowed for in calibrating the actual jack used for stressing.

7.7.3   Duct Friction

Both EC2 and BS 8110 specify the standard formula for calculating the force lost overcoming duct friction when the tendon is stressed. This separates the loss into two parts:

1. That caused by the tendon and duct following the specified profile;
2. That caused by slight variations in the actual line of the duct, which may cause additional points of contact between the tendon and the duct wall.

BS 8110 calculates the losses in terms of the coefficient of friction (μ) and a profile coefficient (K) expressed as a value per metre length. EC2, on the other hand, uses the coefficient of friction (μ) and an unintentional angular displacement per metre length (k). The designer should note that $K = \mu k$.

BS 8110 gives a range of values of μ for lightly rusted strand in different ducts and recommends that K should lie between 17×10^{-4} and 33×10^{-4}. EC2 provides values of μ for a range of tendons which fill about 50 per cent of the duct and suggests that k should be between 0.005 and 0.01 rad/m. Taking both factors together, the EC2 recommendations would generally lead to lower losses. However, it is generally better to refer to literature and test results produced by the manufacturers of prestressing components, and both codes allow the designer to adopt this approach.

7.7.4   Elastic Deformation

The method specified for calculating losses due to the immediate elastic deformation of the concrete when the prestressing force is applied is the same in both codes. However, as the value for the modulus of elasticity specified in EC2 is approximately 15 per cent higher than that in BS 8110 for concrete of the same strength, the losses will be correspondingly lower.

7.7.5   Anchorage Draw-in or Slip

Both codes mention this cause of loss of prestressing force, but neither specifies any values. Appropriate values can be obtained from the anchorage manufacturers and should be checked on site, particularly if the member is short when the loss due to this cause can be critical.

7.7.6   Time-dependent Losses

Time-dependent losses are those due to relaxation of the prestressing tendons, shrinkage and creep of the concrete. The two codes adopt different approaches for calculating these effects. BS 8110 considers each effect separately and the values for each parameter suggested by the code take into account their interaction with one another. EC2 considers the loss due to each effect as if that effect were acting alone and then calculates the combined effect using an interaction formula.

7.7.6.1 Relaxation of Steel

Both codes take as their starting-point the relaxation loss at 1000 hours which is best obtained from the test certificates for the tendons, or from the appropriate material standard. In addition EC2 suggests values for this parameter in the absence of any other sources of data. The 1000-hour relaxation loss is then multiplied by a factor to give the long-term loss. In EC2 this factor is 3, while BS 8110 gives a table of values, all between 1 and 2, which takes account of the tendon type and whether the member is pre- or post-tensioned. The lower values in BS 8110 allow for the effects of stress reductions due to creep and shrinkage of the concrete and, in the case of pretensioning, due to the elastic deformation of the concrete at transfer. The UK NAD specifies that the tabulated values in BS 8110 should be used rather than the blanket value of 3 specified in EC2. While it is acknowledged that a value of 3 is high for low relaxation strand, the values in BS 8110 will underestimate the loss for the reasons just stated. A value of 2 is considered to be more appropriate for low relaxation strand.

7.7.6.2 Shrinkage

Both codes suggest typical values within the main text for the final shrinkage strain of concrete, with more detailed information provided in Part 2 of the British code and Appendix 1 of the Eurocode. The values suggested in BS 8110 are between two and three times smaller than the values given in EC2 and, unless recourse is made to Part 2 of BS 8110, no account is taken of the effect of member thickness.

7.7.6.3 Creep

Both codes suggest typical values within the main text for the final specific creep strain (or creep coefficient) for concrete, with more detailed information provided in Part 2 of the British code and Appendix 1 of the Eurocode. As for shrinkage strains, the main text of EC2 provides information on the effect of member thickness on specific creep strain, while the designer must refer to Part 2 of BS 8110 for this information. The values suggested for the specific creep strain are larger in EC2 than BS 8110, particularly when the age of the concrete at time of loading is less than 28 days.

7.7.6.4 Comparison of Time-dependent Losses

Because of the different approaches to calculating long-term losses in the two codes and the different maximum jacking loads, it is necessary to consider the total effect from the relaxation of the steel, and the shrinkage and creep of the concrete. A comparison of the total long-term losses for a typical prestressed T-beam with varying age at the time of stressing and under different conditions of exposure is given in Table 7.11. The EC2 values have been calculated from the data in the main body of the code, not the appendix. It is not possible to draw general conclusions from a single example, but it would appear that there is reasonable agreement between the different methods for members in an outside environment and where transfer takes place within seven days. For transfer at a later age, EC2 is likely to predict a slightly greater loss than BS 8110. The results are more scattered for an inside environment. The method in Part 2 of BS 8110 will generally give the highest estimate of loss with the EC2 prediction lying between the two values predicted by the British code.

Table 7.11 Comparison of long-term losses (%)

| Environment | Age at transfer | | | | | | | | |
| | 3 days | | | 7 days | | | 28 days | | |
	EC2	BS 8110 (Pt 1)	BS 8110 (Pt 2)	EC2	BS 8110 (Pt 1)	BS 8110 (Pt 2)	EC2	BS 8110 (Pt 1)	BS 8110 (Pt 2)
Inside	20	18	28	19	15	22	17	12	17
Outside	14	14	16	13	11	12	12	9	9

7.8   Anchorage Zones
7.8.1   Pretensioned Members

Both codes provide guidance for estimating the transmission lengths in pretensioned members. EC2 draws the distinction between the transmission length over which the prestressing force is fully transmitted to the concrete; the dispersion length over which the concrete stresses gradually disperse to a distribution which is compatible with plane sections remaining plane; and the anchorage length over which the tendon force at the ultimate limit state is fully transmitted to the concrete. This is a distinction which the designer will find helpful in thinking about the distribution of forces within the anchorage zone.

The suggested values for transmission lengths are compared in Table 7.12,

Table 7.12 Transmission lengths (no. of diameters)

Code	Tendon Type	Concrete grade					
		25/30	30/37	35/45	40/50	45/55	50/60
	Plain or indended	110	99	89	85	81	77
	Crimped	73	66	60	57	54	52
BS 8110	7-wire standard or super						
	strand	44	39	36	34	32	31
	7-wire drawn strand	66	59	54	51	49	46
EC2	Strands/indented wires	75	70	65	60	55	50
	Ribbed wires	55	50	45	40	35	30

where it can be seen that there are significant differences between the two codes, particularly for strand. This will lead to transmission zones being 300−400 mm longer in members designed to the Eurocode with a consequential increase in link reinforcement in the members.

7.8.2   Post-tensioned Members

The design of anchorage zones, or end blocks, in post-tensioned members is a very important area of design, but one which is rarely well treated in codes of practice. BS 8110 and EC2 are no exception. Both codes draw attention to this important area and either outline the principles of the design (EC2) or advise the designer to consult specialist literature (BS 8110).

BS 8110 provides guidance on calculating the bursting forces which occur in the prism behind the anchorage. These are checked at the serviceability limit state for bonded tendons and the ultimate limit state for unbonded tendons. At the serviceability limit state the bursting forces are resisted by reinforcement acting at a design stress of 200 N/mm² which is chosen to limit the widths of any cracks which may occur. At the ultimate limit state, the design stress is the usual $0.87f_y$. The designer is referred to specialist literature in order to consider the effects of overall equilibrium and spalling. Only minimum guidance is given on the detailing of anchorage zones, which is as important as being able to assess the forces acting on them.

EC2, in its present draft, is confusing, clause 4.2.3.5.7 instructing the designer to check bearing stresses in accordance with clause 5.4.8, which specifically excludes prestressing anchorages. The designer's attention is drawn to the need to consider bearing stresses behind the anchorage, overall equilibrium and transverse tensile forces, i.e. bursting and spalling forces. The design is carried out at the ultimate limit state and is based on the characteristic strength of the prestressing tendon, not γ_p times the prestressing force. No mention is made about crack control, this presumably being deemed to be covered in the general detailing requirements. It is suggested that overall equilibrium be checked using a strut and tie model and guidance is given on this and on the angle of dispersion of the prestressing force into the concrete section. This is of benefit to the designer.

7.9.2   Anchorages and Couplers

EC2 requires that not more than 50 per cent of the tendons should be coupled at any one cross-section. There is no similar restriction in BS 8110 and it is generally accepted UK practice to couple all the tendons at one cross-section. The reason for the EC2 requirement is that when a coupled tendon is stressed the force between the previously stressed concrete and the anchorage decreases and the local deformation of this concrete is reduced. Increased compressive forces are induced adjacent to the anchorage and balancing tensile forces between adjacent anchorages. Providing uncoupled tendons across the joint reduces this effect as well as assisting in carrying the induced tensile forces. Alternatively, it is suggested that such forces could be carried by properly detailed passive reinforcement.

7.9.3   Minimum Area of Tendons

Both codes contain requirements regarding the minimum area of prestressing tendons. The BS 8110 requirement expresses this in terms of the ultimate moment of resistance of the section which must be greater than the moment required to produce a tensile stress in the extreme fibre of $0.6\sqrt{f_{cu}}$, taking the prestressing force as its value after all losses. EC2 uses the same requirement as for a normally reinforced concrete member, with f_{yk} substituted by f_{pk}. In all practical cases this means that EC2 will require a minimum area of longitudinal reinforcement of 0.15 per cent, which could be made up of normal reinforcement or prestressing tendons with no adjustment made for the different characteristic strengths.

7.9.4   Tendon Profiles

The design for shear in EC2 leads to a requirement for longitudinal reinforcement in addition to that required for bending. This reinforcement is generally allowed for by a horizontal displacement of the bending moment envelope (the 'shift' rule). When the flexural design of a prestressed concrete member is governed by the ultimate limit state, which could be the case when the design criterion is a limiting crack width of 0.2 mm (class 3), then the longitudinal reinforcement will need to be increased or the tendon profile adjusted in order to meet the shear requirements.

7.9.5   Shear Reinforcement
7.9.5.1   Minimum Area
In BS 8110 the minimum area of shear reinforcement depends only on its yield strength, while in EC2 it also depends on the concrete strength. The requirements are summarized in Table 7.14, where it can be seen that EC2 will always require a greater area (concrete grades 12/15 and 20/25 are not allowed for prestresseed concrete members).

On the other hand, no guidance is given on how to design against spalling bursting forces. The Eurocode provides more guidance on the detailin anchorage zones than the British code, but the information given could be precise.

Both codes draw the designer's attention to the importance of paying s[attention to anchorage zones which have a cross-section different in shape that of the general cross-section of the member.

For more information on the design and detailing of anchorage zo[prestressed concrete, the designer is referred to the excellent documents pr[by CIRIA and the Institution of Structural Engineers (see Bibliography]

7.9 Detailing

The detailing provisions given in section 5 of EC2 apply to both prestres normal reinforced concrete. In this section only those provisions which spe relate to prestressed concrete are discussed. A more general compa[detailing requirements is given in Chapter 6.

7.9.1 Spacing of Tendons and Ducts

Both codes specify the minimum allowable clear spacing between adjacen or ducts. The requirements are broadly similar, generally being the g the tendon or duct diameter, or the aggregate size plus 5 mm. EC2 also absolute minimum clear spacings, but these are unlikely to be critical whe is taken of the need to be able to compact the concrete around the ten[the normal range of tendon sizes, BS 8110 would allow slightly small[spacings in pretensioned members, but again this is not likely to] significance in practice. For post-tensioned members, it should be note[bases its requirements on the outside dimensions of the duct, while BS the internal dimensions. The requirements of the two codes are c[Table 7.13.

EC2 does not specify any additional requirements for the minimum cl in the plane of curvature between curved tendons. This is a serious (very high local forces can be developed. The designer is recomme[the requirements given in BS 8110.

Table 7.13 Comparison of minimum clear spacing requirements for pr[tendons and ducts

		EC2			
Pretensioned	Vertically	$\geq d_g$	$\geq \phi$ ≥ 10 mm	$\geq$	
	Horizontally	$\geq d_g + 5$ mm	$\geq \phi$ ≥ 20 mm	$\geq$	
Post-tensioned	Vertically		$\geq \phi$ ≥ 50 mm	$\geq$	
	Horizontally		$\geq \phi$ ≥ 40 mm	$\geq$	

Note: d_g = aggregate size. ϕ = diameter of tendon or duct, as appro[

Table 7.14 Comparison of minimum shear reinforcement ratios ($A_{sw}/b_w s$)

Concrete grade	Steel grade				
	EC2			BS 8110	
	S220	S400	S500	$f_y = 250$	$f_y = 460$
12/15, 20/25	0.0016	0.0009	0.0007		
25/30 to 35/45	0.0024	0.0013	0.0011	0.0018	0.0010
40/50 to 50/60	0.0030	0.0016	0.0013		

7.9.5.2 *Maximum Spacing*

In both EC2 and BS 8110 the maximum longitudinal spacing of shear reinforcement depends on the magnitude of the applied shear force. In BS 8110, the shear force is compared with the design ultimate shear resistance of the concrete (V_c), while in EC2 it is compared with the maximum design shear force which can be carried without crushing the notional concrete compressive struts (V_{Rd2}). As the applied shear force increases the maximum allowable spacing decreases. In addition, EC2 specifies absolute maximum spacings which are such that this code is likely to be the more onerous.

In the transverse direction, the maximum spacing specified by EC2 is again a function of V_{Rd2}, while BS 8110 specifies that the spacing should not exceed the effective depth (d), whatever the magnitude of the applied force. Again EC2 is the more onerous, particularly so for applied shears greater than $0.2V_{Rd2}$.

Applying the EC2 rules will result in the adoption of smaller diameter shear links at closer centres. This will be beneficial in controlling any shear cracking which occurs and may, in any case, be required to satisfy the EC2 crack control requirements.

Bibliography

CEB–FIP Model Code 1990 Bulletin d'information No. 195, Comité Euro-international du Beton, Lausanne

CIRIA Guide 1 *A Guide to the Design of Anchor Blocks for Post-tensioned Prestressed Concrete Members*, Construction Industry Research and Information Association, London 1976

CP110 : Part 1 : 1972 *Code of Practice for the Structural Use of Concrete*. British Standards Institution, London

DD ENV 206 : 1992 *Concrete, Performance, Production, Placing and Compliance Criteria*. British Standards Institution, London

Standard Method of Detailing Structural Concrete. The Institution of Structural Engineers, London 1989

Index